THE ICE-BOUND WHALERS

THE ICE-BOUND WHALERS

The Story of the *Dee* and the *Grenville Bay,* 1836-37

Edited by James A. Troup

Maps by Anne Leith Brundle

1987
THE ORKNEY PRESS
in association with
STROMNESS MUSEUM

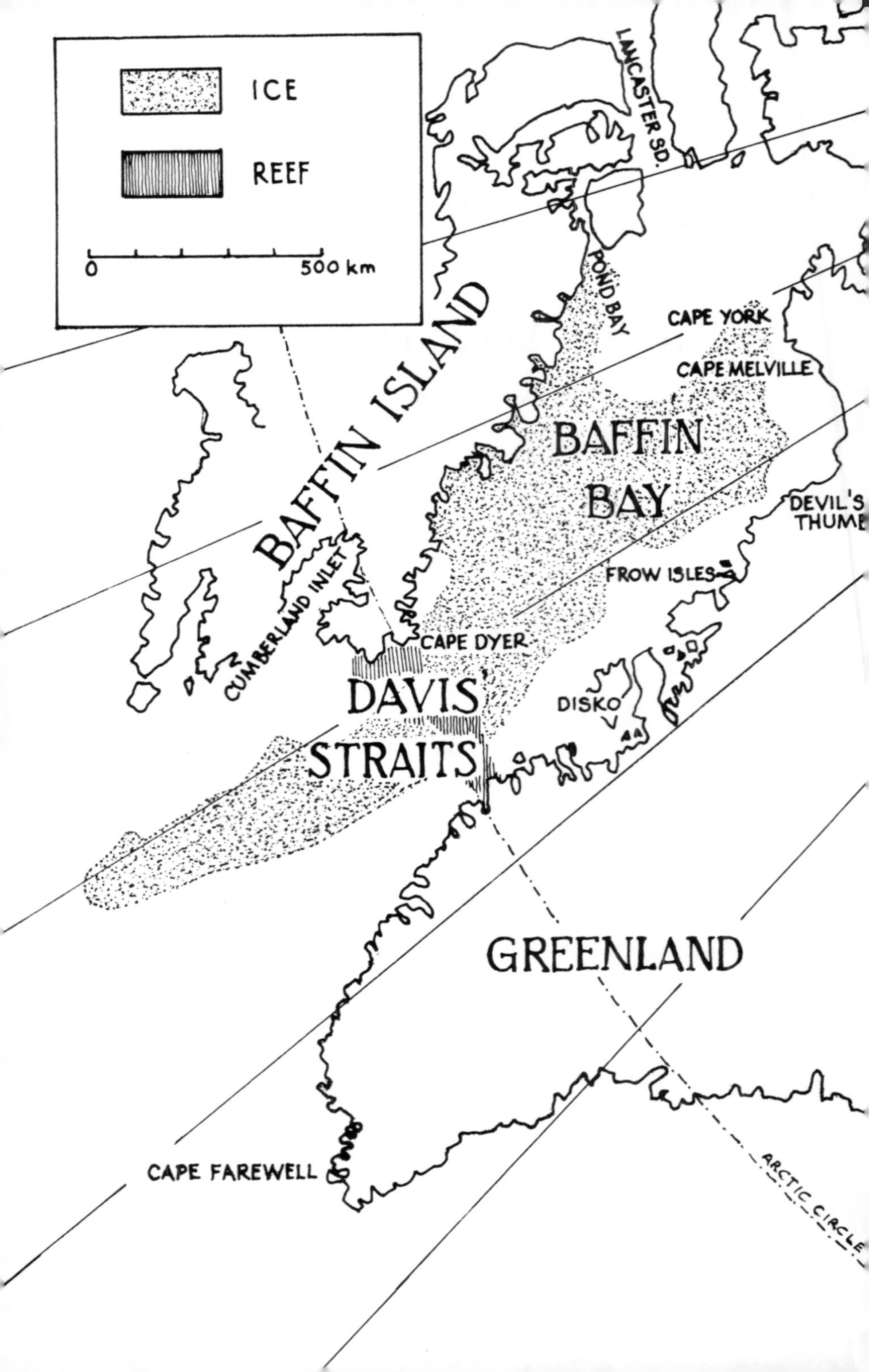
ICE
REEF
0
500 km
LANCASTER SD.
POND BAY
CAPE YORK
CAPE MELVILLE
BAFFIN ISLAND
BAFFIN BAY
DEVIL'S THUMB
FROW ISLES
CUMBERLAND INLET
CAPE DYER
DAVIS STRAITS
DISKO
GREENLAND
CAPE FAREWELL
ARCTIC CIRCLE

To the whaling seamen who suffered
and died in the Arctic ice-fields,
this book is dedicated.

During the late spring and the summer of 1837, whaling ships crept one by one into Stromness harbour in Orkney, manned by the crippled remnants of their crews after a winter locked in the Arctic ice — a disastrous season from which the British whaling industry never recovered.

Towards the end of 1837 there was held the inaugural meeting of the Orkney Natural History Society, laying the foundations of a Museum in Stromness which since that time has built up collections on the natural and maritime history of Orkney, including the whaling industry which regularly enlisted crews from Orkney and Shetland.

The 150th anniversary of both events is marked by this publication, which contains a history of the Museum and eye-witness accounts of the ice-bound whalers by crew members of the *Dee* and the *Grenville Bay*.

Published 1987 by
The Orkney Press Ltd.,
12 Craigiefield Park,
Kirkwall, Orkney

ISBN 0 907618 14 6 (Hardback)
0 907618 15 4 (Paperback)

General editing and book design
Bryce S. Wilson

Printed by The Kirkwall Press,
"The Orcadian" Office,
Victoria Street, Kirkwall, Orkney

CONTENTS

ILLUSTRATIONS

SOURCES OF ILLUSTRATIONS

City of Kingston Upon Hull Museums & Art Galleries — *f.p. 29, 36, 76, 93.*

Dundee Museums & Art Galleries — *Whaling Scene by John Gowland.*

Glenbow Museum, Calgary, Alberta, Canada — *cover and frontispiece, f.p. 92.*

Orkney Library & Archive, Kirkwall — *f.p. 12 (top) and 13.*

Brown, Son & Ferguson, Glasgow — "The Whale."

"The Orcadian," Kirkwall — Adam Flett.

Aberdeen Public Library — *p. 45, 46, 47.*

Peter K. I. Leith — Whaler's pay warrant.

Trevor Housby — Whaling harpooner in Azores (permisson sought).

John Sinclair — William Wilson.

Harry Flett — *p. 40, 44.*

John Brundle — St Michael's Kirk, Harray.

James Kirkness — Mrs J. J. M. Firth.

The remainder are from the collection of Stromness Museum.

ACKNOWLEDGEMENTS

We also acknowledge the assistance of Orkney Library and Archive, Dundee Public Libraries, Montrose Museum, Cumbria Archive Service, The Divisional Library, Wick, Newcastle Upon Tyne Public Library, Greenwich Maritime Museum, Shetland Archives, the Flett family, Dr Derek D. Johnstone, Mrs C. Stamp, Mr R. Shearer, and especially: Aberdeen Public Library, Orkney Museums Service, David S. Henderson of Dundee Museums, Arthur G. Credland of Town Docks Museum, Hull.

Whaling scene by John Gowlan

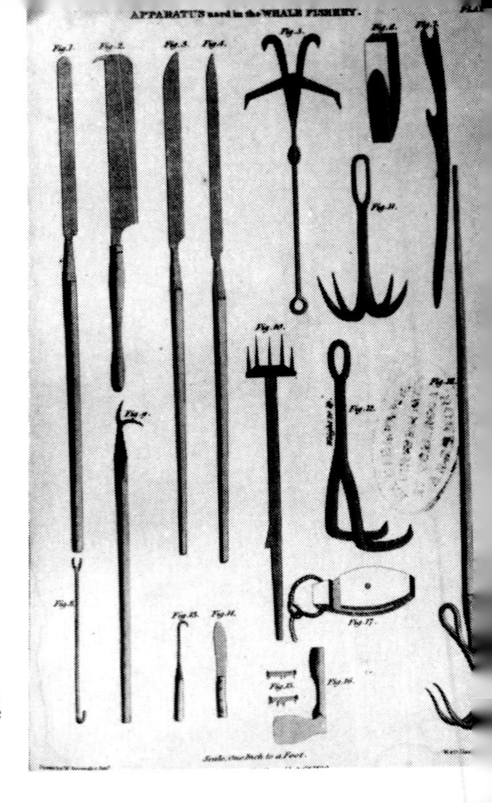

Apparatus used in the whaling industry, including flensing knives and ice grapnels.

Whaling harpooner in the Azores, 1969.

From *Sea Songs and Shanties* collected by Captain W. B. Whall.

The Whale.

THE following is a genuine sea song, and was a prime favourite fifty years or so ago among the old-fashioned sailors.

Stromness in the mid 19th century.

With tarry dress we reached Stromness
Where we did go ashore,
With whalemen so scarce and the water even less,

Orkney and Arctic Whaling

JAMES A. TROUP

The Flotta Flare shining brilliantly across the land-locked waters of Scapa Flow stands as a symbol of the importance of North Sea oil to Orkney today, her employment and her economy. The great fighting ships that used this huge natural naval base in two world wars have given place to giant tankers, playing follow-my-leader to and from the single-point moorings off Flotta. Mineral oil, a vital fossil fuel for twentieth century man has turned Orkney into one of Britain's biggest ports, well above Liverpool or the Clyde. In tonnage handled, Orkney stands eighth on the league table of ports in Britain, third in Scotland, though dwarfed by the oil metropolis of Sullom Voe in Shetland, now Britain's leading seaport.[1]

It is not the first time that oil has brought prosperity to these islands. Before petroleum was discovered men utilised other oils — vegetable and animal — to meet their needs, as a softening fluid for coarse hard fabrics, as a lubricant for machines, as the source of light in homes and, sometimes, in town streets. The most important source of oil for all these purposes was the blubber of the whale. As in Faroe today, schools of caa'ing whales would be driven into shallow water and be slaughtered for their meat and fat.

Visitors to the islands recorded such events in considerable detail due to their novelty value.

> "On Saturday morning last a shoal of about 200 of a small species of whale paid us a visit, they were no sooner seen than almost every boat in the town was manned & a hot persuit commenced. After following them about for an hour they

succeeded by shouting & firing balls amongst them in frightening them into shallow water & driving them ashore just at the back of Stromness harbour. Here every boat's crew kills & secures for itself as many as possible & the moment the fish are ashore or in very shallow water a dash is made amongst them, the Men jump out into the water, harpoons, lances, knives &c are brought into play, & such a scene as it is of blood butchery & confusion can hardly be imagined — the water for a great way about is soon died red with blood & the fish in the agonies of death lash it up to a great height with their tails, it is necessary then to keep clear of them for they sometimes stave a boat & the blows that will do that would certainly break a man's back or seriously injure him. These fish went ashore at a bad time that is with flowing water so that only about 120 of them were captured & nearly as many more got away. The largest of them measured nearly 20 feet in length . . . Nothing here is so eagerly engaged in by all classes as a Whale hunt, because if successful there is profit as well as sport. One of the largest of the whales lately captured I was told yielded 13¾cwt of blubber, worth here about 14s. a cwt, & each cwt is calculated to yield about 8 or 9 gallons of oil. Last year about this time 195 of these fish made their appearance off Flota an Island about 10 miles from hence where they were driven ashore & everyone of them secured — there it is the custom for all persons engaged in driving them ashore & slaughtering them to share equally, so that these 195 fish were sold by auction & fetched £500.12.6 a sum of money very acceptable to these poor Islanders — after a whale visit many a house is enabled to burn a lamp that would otherwise have had no other light thro' the long winter evenings than what a turf fire would have afforded. These fish seldom make their appearance here but once in a season & are not always captured — two years ago a large parcel of them

Charles Peel, man-o'-war's man, whaler and pedlar, photographed in Stromness at the turn of the century.

William Wilson, a native of Graemsay and a veteran whaler, who sailed in 1870 on the *Intrepid* of Dundee, the last whaler to ship part of her crew from Stromness.

came up here & 70 or 80 boats persued them about the whole day but we never could get them out of the deep water so that not one of them was taken."[2]

Such exciting whale hunts occurred but rarely. However valuable they were to small poor communities they could do nothing to meet the nation's needs. Pilot whales are toothed and have little blubber. They were ignored by the distant water whaling fleets. Very different was the giant of Arctic seas — the Greenland right whale.

The Greenland Whale

Hunted at Spitzbergen in the early seventeenth century the Greenland whale had been pursued with varying success ever since, along the ice edge from Spitzbergen to Jan Mayen Island and so to the east side of Greenland. When Greenland waters proved less lucrative attention was directed to the waters to the west, between Canada and Greenland. It was a classic tale of over-exploitation.

The Greenland whale was the whaleman's Right Whale. It was the right one for many reasons. Not only was it a docile, slow moving giant up to 60 feet long, not only did it possess a layer of blubber 12 to 18 inches thick, but also, when killed, the creature floated on the surface, whereas the sleeker, swifter rorquals sank and would be lost. This type of whale was very slow to associate men with danger and so was relatively easy to approach in rowing whaleboats; yet it was timid and easily startled by the plash of an oar or even by a bird landing on its back.[3] The gentleness of the whale did not mean that harpooning it was simple and safe. A badly positioned whaleboat might be upset or shattered by powerful tail flukes, as broad as a whaleboat was long, jerking in response to harpoon-induced pain. "Always strive to avoid their tayles (because with that part they strike and if they hitt a boate will break it in pieces) but if you bear up to their head and foreparts, then are you more secure."[4] That advice, given by Lancelott Anderson in the seventeenth century, was valid for any period of hand harpooning of the Greenland whale.

The Dundee whaler *Scotia* in Stromness harbour in the early 20th century.

Whaling mishap from Scoresby's *Account of the Arctic Regions.*

And when we reached that whale, my boys,
He lashed out with his tail,
And we lost a boat, and seven good men,
And we never caught that whale — brave boys,
And we never caught that whale.
from 'The Whale'

Once the prey was slain, there followed the shoulder-tearing task of bringing back the carcase. "Rowing back to the ship while towing a seventy barrel whale looked impossible; the oar seemed to dig into the same little whirlpool of water it had just left."[5]

The long hours of Arctic daylight meant that almost unbelievable hours could be worked when necessary. A large whale secured to the side of the ship had an unbalancing effect that could make her almost unmanageable with a rising gale. Besides that, man was not the only predator. A dead whale would soon attract a myriad of gulls and, more seriously, hungry sharks as avid as the whalemen to strip the bones. They would gain more than the men with much less effort for, in the end, the carcase would be cut loose with perhaps 20 or 30 tons of meat adhering to the bones. The men desired only the fat from the body and the whalebone from the mouth.

Rest was lacking when a whale was alongside. A captain would allow just as long as he felt necessary to keep his men going. William Barron tells a story of killing a whale at about 3 p.m. "About two in the morning we got alongside having had nine miles to tow" (with help from the other boats). "We were all ready for a rest, and some of us were very wet the whole time we had been employed in killing the whale. After securing it to the ship, we had a good four hours' sleep. Flensing followed."[6] If the whalemen had delayed too long the sharks would have done severe damage to the underside of the whale.

Wearing long metal spurs (spikes like the crampons of a mountaineer climbing on snow and ice) the harpooners walked on the body of the whale, making cuts through skin and blubber with razor-sharp flensing knives and blubber spades. Good balance was essential for one slip might lead to being "crushed like a peanut shell between the rolling carcass and the side of the ship."[7] That gruesome prospect was rather less likely in the Arctic than in the open ocean of which Whipple was writing, for the ice tended to maintain considerable areas of calm water.[8] In Davis Strait when a "fish" was captured it was worth sailing several miles to lie in a bay "until the troublesome process of making-off was accomplished."[9] When all was ready the harpooners withdrew to the accompanying whaleboats and the blubber was ripped in sheets by the pulleys aboard ship and dropped on the deck. The carcase was then laboriously turned to expose the next face of blubber.

Aboard ship the blubber was cut up into small pieces and packed into barrels through the bung hole. Eventually, in a rancid state, it would reach the home port to be boiled till it yielded the train oil.

The right whale feeds by sucking in a mouthful of seawater and expelling it through the fringed plates of baleen that hang from its upper jaw, catching upon them the tiny shrimp-like krill that provide its food. The open mouth of a big Greenland whale could contain a modern kitchen. The longest of the 500 or 600 pieces of whalebone could be 12 or 13 feet. Whalebone was a bonus — an extra source of income from a whaling trip and usually a valuable one. Tough, flexible and springy, it had a multiplicity of uses. Its price was best when fashion decreed that the female form should be tightly confined. Then it was in great demand for stays and corsets; but it had many other uses — from fishing rods to brooms to sweep the stable yard, from coach springs to rowlocks, from umbrella ribs to sofa stuffing. A large whale could be expected to yield one to one and a half tons of this valuable commodity.[10]

The Politics of Whaling

England had pioneered seventeenth century Spitzbergen whaling; Scotland had followed later on a much smaller scale. Both had been miserable failures who saw all the profits going to the Dutch. "For almost a century English whaling was a dismal affair conducted on a level akin to an Englishman with a thimble emptying the same tun as a Dutchman with a bucket."[11] As a result Britain imported what whale oil she needed.

As the eighteenth century advanced, mercantilist policies were followed with more rigour. From 1733 the government offered a tonnage bounty, in part to encourage whale catching so that oil imports would be reduced, but more with the aim of increasing the pool of available seamen in time of war. Not till 1750 did these subsidies produce a notable effect. At that time a bounty of £2 per vessel ton coincided with a rising demand for whale products from developing industry. Growth was not consistent. In the Seven Years' War (1756-1763) and to a more devastating extent in the War of the American Revolution (1775-1783) the press gang disregarded passes that gave exemption from conscription to essential men on the whaling vessels. Many shipowners found it simpler to take assured profits by hiring their whaleships to the state as transports.

American independence gave a great and continuing boost to the British whaling industry. Not only were the Dutch excluded by crippling taxes from the British market, but so were the traders of the infant U.S.A. British whaling companies now had a clear field to supply a buoyant home market.[12]

Davis Strait and Baffin Bay

The first venturers into Davis Strait were intent on a safe fishery south of the ice barrier, but as whales became harder to find there it became necessary to move further north. The course lay through the seldom frozen coastal waters as far as Disko Island. To the west lay 'the reef', a ridge that runs across Baffin Bay. On it icebergs frequently ground and may lie for years. Behind them an ice sheet builds up. It had to be bypassed on the outward journey, but coming home it was usually possible to sail down the middle of Baffin Bay. Just north of Disko it was often necessary to wait for the ice to melt. As leads of clear water showed, ships would press north towards the open bight of Melville Bay, the dreaded 'ship breakers' yard'. Every effort was made to get through quickly. If becalmed, captains would rely on muscle power to keep moving, either tracking or towing. The former involves the crew, wearing special harness, walking on the ice and hauling their ship forward; the latter, pulling on whaleboat oars to the same purpose. Both were exhausting and dreary occupations.

Beyond Melville Bay lay the open sea of the 'west water'. In inlets of Baffin Island and especially in Pond's Bay a rich harvest might be found "and a full ship certain, that is to say if we get there previous to the fifth or sixth of July, the time when the whales pay their annual visit, in route for the north."[13]

Ships and Men

To enter this wild and hostile environment needed special ships and special men. The men had to be hardy and resilient, able to cope with wildly unpredictable

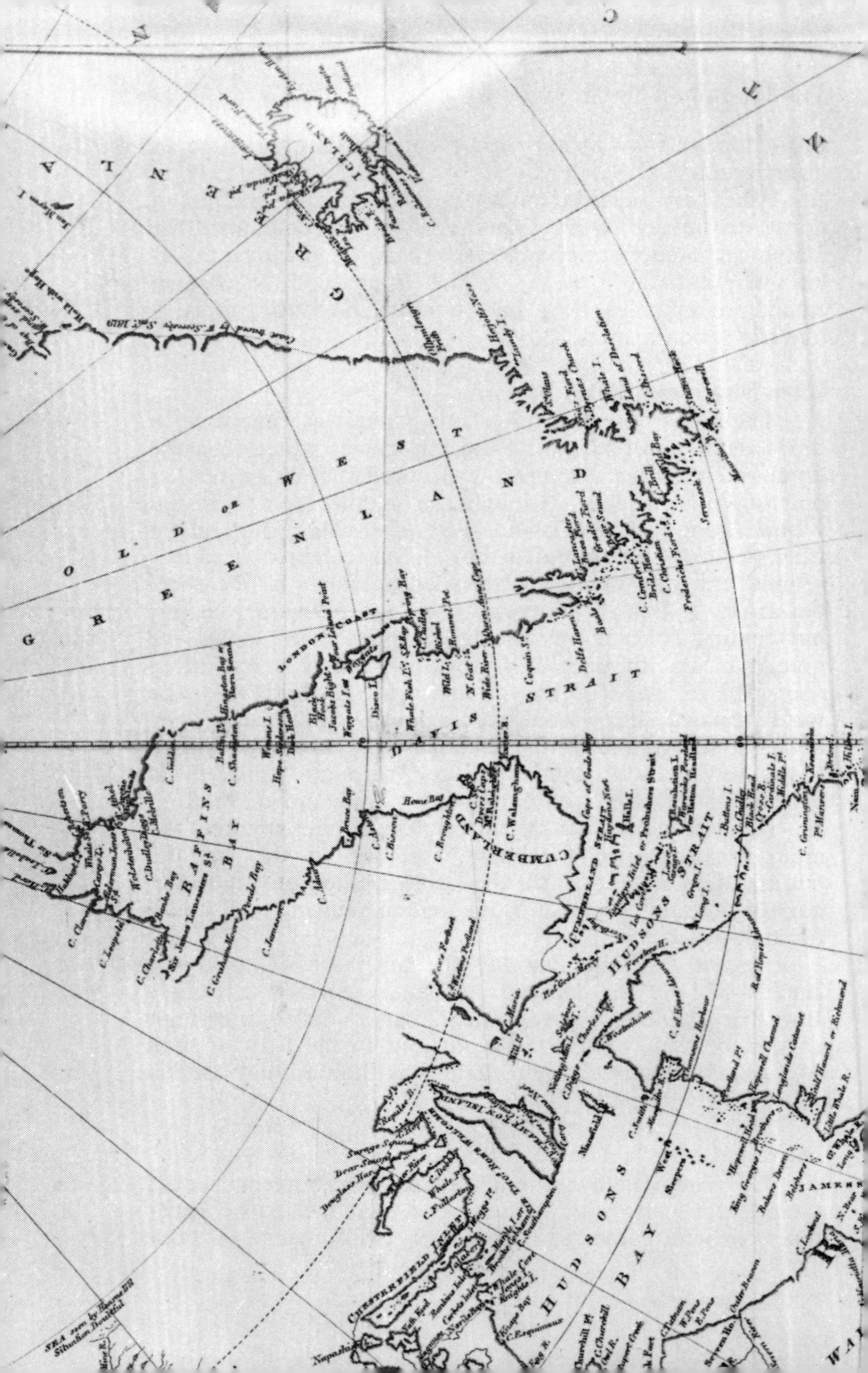

ICELAND
OLD OR WEST GREENLAND
LONDON COAST
DAVIS STRAIT
BAFFINS BAY
CUMBERLAND
CUMBERLAND STRAIT
HUDSONS STRAIT
HUDSONS BAY
CHESTERFIELD INLET
JAMES

mealtimes and with periods of fragmented sleep. On the hunting grounds the whale took precedence over everything else. If called from their bunks they came "running . . . with their eyes scarcely open some of them and their clothes in a sling to put on when they got more time."[14] It did not pay to be a worrier. Unexcitable phlegmatic characters were best for the trade. Special qualities were required of the skipper, above all, knowledge of the area and skill in navigation. When the crow's nest was sent up it was only in part to spy out whales. Much more important was to seek the best route by his interpretation of ice movement. If it were true that "The Lives and Fortunes of Seafaring Persons, in great Measure depend on the Accuracy of their Charts"[15] then this was no area in which to sail. There were no charts. Barron emphasised the need to instruct apprentices on the features of the land as a reminder of sunken rocks and shoals.[16] Whaling high above the Arctic Circle was always a gamble. An experienced and capable captain could lessen the odds.

The crow's nest enabled the captain to show himself a strict disciplinarian. When the wind was light, everything said on deck drifted up to his ears. He was likely to respond with some form of 'make-work', probably sending boats "away for hours when nothing has been seen, especially when it was a cold, bitter day."[17]

Ships faced stresses as well as men. In the age of sail many vessels met shipwreck in the ordinary trading waters of Europe; the Arctic offered the extra danger of an ice nip that might crush the stoutest boards. As a result whaling vessels were specially protected, whether purpose-built or converted merchantmen. Inside, sturdy beams were inserted to resist the pressure of ice on the ship wall which was strengthened by "An additional thickness of oak plank applied on the outside of the ship from the wales down to the water mark from one end of the ship completely to the other."[18]

This technique of doubling (which can be seen in the model of the *Swan,* a converted gun brig) emphasised the stocky breadth of whaling ships and contributed to their general slowness, but, if slow, they were easily manoeuvrable having rigging adjusted to allow a skeleton

Map from Scoresby's *Account of the Arctic Regions.*

crew to manage the vessel amid ice when most men might be away all day in whaleboats.[19] The strength of the vessel might reassure an inexperienced medical student, looking on a stint as ship's surgeon as a fine jaunt, and let him see the romance of drift ice when being wakened by "the bump, bump of the floating pieces against the side of the ship, and [going] on deck to see the whole sea covered with them to the horizon."[20] That was in the open sea on a venture that was primarily sealing and therefore not subject to the same hazards. A more sensitive young man, aboard the *Thomas* of Dundee two years before she was crushed by ice had a much clearer appreciation of the irresistible power of grinding ice after seeing its effect on a wrecked ship. "I am almost afraid to go to my bed after viewing the state of this vessel. Her masts were torn from the deck and crushed upon the ice, her deck was torn asunder and every part that would offer resistance was crushed to atoms."[21] Small wonder that Scoresby confessed to "perpetual anxiety."[22]

By experience it had been discovered that the most suitable size for a whaleship was around 300 to 350 tons. Both the *Grenville Bay* and the *Dee* were of this size.[23]

The Orcadian Connection

How was it that Orkney became involved in this great industry whose operations took place so far away? Whaling from six man whaleboats was so labour intensive that it was natural, when the supply of men nearer home dried up or became too expensive, to pick up extra crew at convenient stopping points. American South Sea whalers, heading for the S. Atlantic and Pacific, travelled much further to crew in the Azores. Shetlanders and Orcadians could pull on an oar and had the additional merit of being cheap. It was that point that Mrs Christian Robertson, a Stromness merchant and whaling company agent used in trying to widen her clientele. "That you may have some idea of the rate of wages here last fishing season I shall hereto subjoin a note of the wages of the ship *Alexander* of Aberdeen and *Cumbrian* of Hull from which you will perceive that the wages are (I presume) much lower than they are in your quarter."[24]

It may be that the sharp increase in the whaling fleet in the mid 1750s could not be met in the immediate area of a port like Hull and that some recruiting was done in the isles. If so, it would be for a very short time, for the fleet declined again as fast as it rose. Any such recruitment would affect Shetland rather than Orkney, for Shetland lay on the route to Jan Mayen Island and the Greenland fishery whereas Orkney would require a detour to the west, a detour in dangerous waters before there were any lighthouses. Hance Smith certainly feels that Shetland was involved thus early. Evidence is thin and some can be interpreted in more than one way.When a group of Shetland lairds drafted a memorial to the Earl of Morton about the pressgang, they were seeking to protect their tenants and the success of the haaf fishery, that is their own commercial interests. Their case was not against the seizure of men, but for its retiming. "And the Season for doing it with the greatest Probability of Success is about the Time, or a little before the Greenland Ships arrive on this Coast which is about the Beginning of March, At which Time should a Tender arrive, it is to be hoped that with the Aid of her Crew, Fourty or Fifty Hands might be secured."[25] This could mean that they were offering up young Shetlanders, who defied the landlord's discipline, to navy service instead of whaling service; it could also be an invitation to authority to ignore whaling exemption certificates and impress whalemen when they were well out of the public eye. Either way, stay-at-home Shetlanders on fishing tenures could go happily about their business.

If Shetlanders were being taken by whaleships at this time it was contrary to the provisions of the Bounty Acts. Not till 1806 was the law changed.

> "Whereas it may be difficult, in the present Circumstances, for Masters or Owners of Ships . . . to be provided with their full Complement of Men at the Ports . . . That it shall be lawful . . . to proceed . . . to Lerwick in the Isle of Shetland, or Kirkwall in the Orkneys, and complete the Number of Men there."[26]

On the other hand, if vessels were shorthanded due to

having men pressed they had little alternative to taking extra crew wherever they could.

Certainly the memorial of 1756 proves that whaling ships appeared regularly off Shetland, probably to buy "such trifling stores as are furnished at a cheap rate in these islands."[27] Acquaintance with the place would have left an impression of deep poverty and of a pool of willing labour.

Once the fishery had opened up Davis Strait, Orkney would come into its own. Its much richer farmland could supply more fresh meat then Shetland just as cheaply. There might be a time lag between arrival in Stromness and delivery of the diminutive cattle produced by the unimproved farms, but seldom did ships need to sail without.[28]. A few days out, fresh meat hanging high in the rigging would freeze solid. Only till then was a periodic dousing with seawater required.

The same signs of unemployment or underemployment of males would have been evident in Stromness, whose inhabitants "are of the lower class of people, who come hither from the other Parts of the Country, and subsist chiefly by means of the ships that rendezvous at their Port. To the Crews of these ships, they sell the common necessaries of Life, which they bring from the other parts of the country."[29] Small wonder that Lieut. Chappell should contemptuously dismiss it as "an irregular assemblage of dirty huts, with here and there a decent house. There is scarcely any thing deserving the name of a street in the place, although it is said to contain a population of two thousand souls. A few years ago it did not contain above one third that number."[30] That he was too dismissive is clear from the number of well-built eighteenth century dwellings in or near the tortuous street, but visitors frequently record what they expect to see. Chappell does, however, note the sharp increase in Stromness' population due to a wartime boom, though greatly exaggerating. Extra hands abounded for the whale ships.

By the time of the (Old) Statistical Account it is crystal clear that the practice of picking up men in the northern isles was well established.[31] Possibly the relaxation that allowed recruitment of Shetlanders late in the

American War of Independence proved impossible to end.[32] Yet, as late as 1793, three years after the first volume of the Statistical Account, the Board of Trade was still trying to prevent the practice.[33]

Once the recruitment of islesmen was established it continued as a regular thing with merchants in Shetland and in Orkney seeking the profits connected with acting as local agent for southern whaling companies. In a place where capital was scarce[34] this was a sensible, almost riskfree, enterprise. Men and stores could be supplied, wages and first payment of oil-money advanced and commission claimed.

James Fea actively advocated developing an oil industry in Orkney, his ideal site being Flotta, because of a sheltered bay and great stocks of peat. Other advantages would be that "almost all our poor people are bred to the sea" and "our people . . . would be glad to go for less wages . . . [and] would be content with much coarser fare, than those that are employed at present."[35] Impractical? But good negative evidence. He makes no mention of whaleships taking on crew, unless in the mysterious reference to taking people for part of the year as "slaves" at starvation wages; but that is much more likely to refer to kelp.[36]

Once only was there any sign of Fea's ideas being put into practice.

In 1813 the Orkney Whale Fishing Company was set up by a group of Kirkwall businessmen including "James Spence, residing in Kirkwall, factor for John Balfour of Trenabay Esquire" — one of the leading lairds. He was one of three James Spences, the most vital to the company being the one who captained the whaleship *Ellen,* purchased from Leith. She was fairly successfully operated, bringing an annual load of blubber to Kirkwall where the firm erected a boiling house and storage sheds, but in 1822 both ship and yard were offered for sale.[37] In the next year, a New Orkney Whale Fishing Company was constituted.

Throughout this period all official documents were signed by James Drever, a Kirkwall merchant. He was to remain keenly involved, being the ultimate purchaser when the new company too gave up business. Drever had continued as an important member of the New Orkney Whale Fishing Co. — the majority Orkney shareholder when it was wound up — but the terms of sale to him suggest that the controlling interest lay with two Leith merchants, James Wyld and John Gavin. That might help explain a seeming switch to Stromness as the sailing point of the vessel. Kirkwall seaman, Francis Burnett, is shown by his certificate of service as making three voyages aboard the *Ellen* (1823-25), each time from Stromness.

By 1829 when James Drever became sole proprietor the value of the Kirkwall premises had fallen from £700 to £300, doubtless reflecting the problems facing the trade. It was probably at this stage that the *Ellen* was disposed of and the 'oily house' used on a much smaller scale for rendering down the blubber of caa'ing whales.[38]

As, year by year, the ships garnered their harvest of recruits in the isles, young surgeons sneered patronisingly in their journals. When experienced captains too offer very sharp comment about the islesmen it needs to be given greater credence.[39] But the ships kept coming. That in itself makes its own commentary; and not all were critical. Capt. Barron was well pleased with them for he never saw them drunk. Many of his Stromness men had previously served with Hudson's Bay Co. and were "adapted for the work through the many privations they had experienced in the Company's employ."[40]

As the industry grew so did the prospect for Orcadians of gaining employment by going to Stromness. Despite Kirkwall being the Customs head port, Stromness was the natural place for whale ships to congregate. Not only does it lie at the west of the islands, but the bay of Hamnavoe, its natural harbour, is the best sheltered anchorage north of the Cromarty Firth. Other convenient bays were also utilised as recruiting grounds like Widewall and Longhope in the South Isles and Deer Sound where the *Fame* was destroyed by fire in 1823.[41] As ship numbers dropped the demand for boatmen rapidly declined. At the peak recruitment could be as high as 20 or 25 men per ship. In 1824 some 700 men went whaling. Eighteen years later the New Statistical Account put the figure "for the last seven

Greenland Ship

AND

BOIL-HOUSE

FOR SALE.

To be Sold, the Ship **ELLEN**, of ***Kirkwall***, as she at present lies at the Pier of that Place, burden **279** $\frac{17}{94}$ Register tonnage, with her whole Stores; also the **BOIL-HOUSE**, with its Fixtures, Yard, and Coopers' Shade, belonging to the same Concern.

The **ELLEN** and her Stores are in good condition, and with the exception of Provisions for the voyage, may be fitted out either for Davis' Straits or Greenland, at a very small expence.—She is a superior sailer, and well adapted for the ***Whale Fishing.***

The **BOIL-HOUSE** is very contiguous to the Harbour of Kirkwall, and is fitted up on an improved plan, with Copper Boiler, and two Leaden Cisterns, each of which will contain about **35** tons of Oil.

Offers for the Vessel, Stores and Boil-house, either separately or together, may be addressed to Mr **JAMES DREVER**, Merchant, ***Kirkwall***, in whose hands may be seen the Inventory of the Materials and Stores.

If not Sold by Private Bargain, previous to 2d December next, the Premises will be exposed to sale by Public Auction, in the ***Town Hall*** of ***Kirkwall***, at Noon on that day.

Kirkwall, **14*th October*, 1822.**

years, on an average, 292 men annually."[42] Unfortunately that is an almost valueless figure. Ship numbers dropped quite substantially over that period from 76 in 1834 to 19 in 1841. Hull which had been so important to Stromness sent only two ships whaling in 1840.[43] The decline of whaling continued apace. The ports that best fought the depression were those that concentrated on seals and took the odd whale as a bonus. Stromness was not affected by them, but by the purely whaling centres. Whaleships came in from time to time for the rest of the century, but the scale of decline is underlined by the *Orkney Herald* reporting that a third whaler was expected on the heels of the *Truelove* and S.S. *Esquimaux* then crewing in Stromness, a greater number "than for a number of years past."[44] By the early twentieth century only the few remaining Dundee whaleships were calling, but were coming fully manned.

It is difficult to be sure when recruitment was at its greatest in Stromness. It may have been in the boom years from 1813 to 1820 or more likely in the 1820s when the industry as a whole had started on a slow decline, but when there was a sharp switch of vessels from the exhausted Greenland waters to Davis Strait and Baffin Bay.[45] When Mrs Robertson was enthusiastically canvassing for fresh customers it must have seemed a golden summer. She did not have the problem of disposing of the whale oil in a shrinking market. Gas lighting was spreading rapidly in big towns, taking away one of the principal outlets for whale oil. At the same time fine cloth was returning to imported rape oil, a much more suitable material for cleaning the wool, as import duties were drastically cut.

In the season after Mrs Robertson confidently predicted no labour shortage wages rose, suggesting that just such a shortage had arisen. Instead, she blamed the brash conduct of another agent.

> "The wages this year took an uncommon rise for which you are obliged to the agent for Mr. Marshalls, who I am sorry to observe did much injury to the trade. I engaged nearly 400 men this year more than half at 35/- & 1/- per Tun in fact

> the wages never advanced untill ships arrived to his address when it gradually advanced — This Gentleman intends visiting Hull this summer for the purpose of increasing his business but I trust the ship owners will see their Interest better."[46]

It seems to have been a temporary wage rise. By 1830 wages for the better men were as they had been in the early part of the 1825 season. Hardly surprising. Profits were dropping fast.[47]

Disaster of 1830

For the most part the weather in Baffin Bay was open in the 1820s. It may be that "expeditions . . . were approaching the limits of physical endurance,"[48] but they were still blissfully ignorant of what could happen in a bad weather season. They received their first severe lesson in 1830, to be reminded at intervals throughout the decade.

There was nothing untoward about the start of the season, no unusual difficulties as the fleet worked northward. Not until 73° N. did they meet any major ice sheet. There,

> "immense floes of prodigious thickness formed a . . . rampart. At length a breach opened, and the ships advanced by the laborious operation of tracking and towing over another degree with an alluring prospect of open water; but the scene soon altered, boundless fields of enormous solidity formed an impenetrable breastwork, along which was assembled most of the shipping".[49]

The wind had changed. A long period of southerly weather drove floes against the land ice. Ice docks gave scant protection that year. When they collapsed only a ship's own strength remained. No matter how strengthened the hull no ship could long withstand the relentless grinding pressure of ice.

> "At one fell swoop the *Baffin* was cut asunder; in the same dreadful spot the beautiful *Ville de Dieppe* and *Achilles* of Dundee were pressed to the water's edge; the *Ratler,* an old

> sloop of war, that, by her firmness and wedge-like form, had rode triumphant over the floes in 1821, was now projected from the surface and shivered in her fall.
>
> ". . . The pressure came with fearful rapidity, it found the *Hope* of Peterhead in safety, but ere ten minutes were elapsed, the point of her main-top-gallant-mast only appeared . . . When the horizon cleared up after a storm, it was common to look around — many of the fleet could no longer be observed, but boats scattered on the ice, and straggling tents, sadly told what had been lately passing."[50]

Nineteen British whaleships and one French were crushed in Melville Bay; many more limped home badly damaged. For some, still iced up in September, the grim prospect loomed of winter in the Arctic, "provisions being meted out on half allowance, and stores of birds accumulated."[51] However, the ice relented. All escaped. A dire warning had been delivered about the power of the Arctic. Loss of life had been amazingly light, yet many men must have been severely affected by cold and exposure and have been unfit ever again to go to sea. As to that we can only guess; there are no statistics to guide us.

Hospital in Stromness

No other year took such a toll of ships as 1830; no other disaster was so sparing of human life. The early 1830s were drab seasons of ship losses and poor catches. Then, in two successive years — 1835 and 1836 — men and ships were caught behind the ice and held there for much of the winter. Two ships, the *Norfolk* of Berwick and the *Grenville Bay* of Newcastle had the unenviable experience of being trapped in both years. On the first occasion they suffered little but short rations and alarm, and reached home in January. Others were not so lucky.

1835 had been a difficult season. Ice had prevented movement up the Greenland side of the straits beyond

'he Plainstones, Stromness, in the late 19th century.

'he piers of Stromness, with the house used as a hospital for the whalers (right, foreground).

Frow Islands. The fleet had been obliged to retreat and try to work north along the coast of Baffin Island. There, at much the latitude of Disko, nine vessels were beset, most of them completely unprepared for the experience. On 1st October, an officer of the *Viewforth* wrote:—

> "The worst of it is, all the ships are very short of provisions — we are now on one and a-half biscuit a-day, one-half pound of beef, and about half a teacupfull of meal. The cold, too, is intense — the ice on top of my bed being one-fourth of an inch thick. Indeed we cannot walk the deck above half an hour at a time."[52]

A month later the oatmeal was exhausted and the bread ration considerably lessened. When the *Middleton* of Aberdeen was crushed in mid-November they gained extra mouths, but "very little increase of the means of supporting them."[53] Food was not the only shortage. By January, "when any of the men now is needing Physic, they are obliged to drink salt water."[54]

When by early December as many as eleven ships were still missing from their home ports — though for most a reasonably accurate location could be offered — Hull shipowners sought the aid of the Admiralty, which initially, was not forthcoming. The state of public opinion, evinced by petitions from all parts of the country and letters to leading newspapers, quickly made the Admiralty relent, but on conditions. If the Hull shipowners and insurance underwriters would find a ship and volunteer crew "the Admiralty would commission her, fill her with stores and provisions and pay the crew, who would be discharged as soon as the service was performed."[55] In addition, a lot more money had to be raised privately.

On that basis, Sir James Clark Ross, an officer with great Arctic experience, was put in command of the expedition. Given a free choice by the shipowners he picked the *Cove*. He sailed exactly three weeks after he went to Hull, with a fully fitted-out and provisioned ship. By then only six whalers were still fast.

The relief expedition met dreadful weather. A terrifying

:utting an ice dock for the whaler *Dundee,* 1826.

storm off Iceland forced Ross back to Stromness for major repairs. It was a fortunate occurrence, for soon:—

> "Thirty-four seamen, in a dreadful state of suffering . . . landed from the two whale ships, *Jane & Viewforth,* nineteen of them being part of the crew of the ship *Middleton* of Aberdeen, lost in the ice. The greater number are recovering although slowly. From the intense cold amputations were found necessary in some cases."[56]

In collaboration with James Login, Ross set up a temporary hospital in "a large empty house" and put Dr Hamilton in charge.[57] That was the most useful part of the voyage of the *Cove* for she was able to put ashore bedding and antiscorbutics for the sick. "If the *Cove* had not providentially been there the loss of life would have been greater for all Stromness could not have supplied them with medicines and bedding nor would the *prompt* measure have been adopted." Mr Login was essential in this for "there was no other agent present in *sufficient credit* to take such responsibility."[58]

The presence of John Macaulay Hamilton, a son of the manse of Hoy, enabled Sir James to retain his surgeon and two assistant surgeons when he sailed again. They would be greatly under-employed: he made no further contact with the missing ships. Hamilton did — when the *Lady Jane* arrived with only 8 of the 64 men aboard even able to crawl. Since getting clear of the ice she had sailed without adjusting close-reefed topsails, for Captain Leask alone had been fit to go aloft.[59]

From the first, Mrs Humphrey's house had been too small. It could accommodate only 26 or 28 patients, the rest being "distributed amongst their friends or acquaintances."[60] After three weeks of suitable diet some of the original scorbutic patients were likely to have been sufficiently recovered to go home. Even so, the pressure of this new shipload of scurvy-ridden seamen upon the meagre resources of little Stromness must have been great indeed.

The long period of worry about absent relatives gave

way to bitter mourning when the reality of the situation was revealed — of 25 Orcadians who shipped aboard the *Lady Jane* 13 were dead.

> "The cost has fallen hard on Stromness to which no less than 6 belong, there is one Kirkwall man — John Saunders — 1 Orphir man, 2 from Ireland, 1 from Birsay, 1 from Firth & from Rendall 1. — The appearance of the survivors — the distress of the friends of the dead on hearing the first accounts baffles all description. It drew tears from the eyes of many an unconcerned spectator. Such a heart-rending scene Stromness never witnessed before."[61]

Thomas Balfour, M.P. for Orkney and Shetland, gave notice of a bill compelling whaleships to carry 12 months supplies. Giving his support to the idea, Andrew Plummer, one of the owners of the *Lady Jane,* claimed that his company already followed that policy. The real trouble was that the seamen "gave themselves up for lost."[62] Adequate supplies of food would plainly have been a help in keeping up spirits, but neither Balfour nor Plummer seems to have appreciated the need for anti-scorbutics. Quality was important, too. Surgeon Wanless was sharply critical, but only to his private journal, of the way that "poor sailors are imposed upon by every person in whatever line of business they deal almost."[63]

'Grenville Bay' and 'Dee' 1836-7

No doubt ailing whalemen were visited by friends passing north en route for Baffin Bay; no doubt those friends left with the heartfelt wish that they should not be subjected to such trauma. Surely the run of ill luck must end! It was not to be as is amply demonstrated by the two texts reprinted in this volume. A number of closely related themes dominate — cold, hunger, isolation, danger and disease — but admiration of natural beauty breaks through occasionally. As time passed and scurvy tightened its grip, deep gloom characterises the journals. More and more shipboard friends seemed destined for a watery grave.

Periodically religious faith peeps through, but never does either account become a religious tract as did the published account of the *Jane* and *Viewforth* in the previous year.

Despite long periods of delay due to ice conditions the journals make little of the outward journey. No hint that the captain had made a desperate gamble for the sake of financial reward. No-one aboard desired to return home with only monthly wages to claim — no-one desired a 'clean' ship. Many ships had a bad season. Of 10 Peterhead ships in Davis Strait 5 returned 'clean'; the others caught a total of 5 right whales and 2 bottlenose whales; only the *Eclipse* with 3 whales giving 50 tons of oil would have made a profit, and an adequate income to keep ordinary seamen through winter.[64]

Yet there was scant benefit in staying on to catch whales if you failed to leave the Arctic. The gap could be wafer-thin. In 1835, when the two ships were close together, the effect of wind on ice was to release the *Grenville Bay* but imprison tightly the *Lady Jane*.[65] In 1836 there was nothing so dramatic, but the time gap must have been slight between the *Friendship* escaping south and the *Dee* and her companions, the *Thomas* and the *Advice* of Dundee, the *Norfolk* of Berwick and the *Grenville Bay* of Newcastle, being trapped for the winter.

Once beset, it was necessary to protect the ships as best they could. Summer in Melville Bay often entailed cutting an ice dock for one, or more often, two ships together. As a winter resource this proved about as valuable as crossing one's fingers. The two related journals show how easily the illusion of security could be shattered. Time and again men had the heavy labour of cutting a dock only to see its walls collapse under pressure of moving floes. The illustration of the *Dundee* shows the tall tripods — whalemen called them 'triangles' — erected on the ice to support the 14 foot ice saws and also indicates something of the pattern of ropes on which the men had been hauling prior to breaking off to push cut pieces out of the dock.

Hard work on inadequate food and with no prospect of drying sodden leather boots, nor any of "his bale of clothing",[66] made illness more likely. Yet, captains were convinced of the need to lay in clean ice from icebergs as raw material for drinking and cooking water, even if it meant miles of pulling boats or sledges, toiling over pressure ridges. Gentle exercise would be beneficial as a regular regimen; tough exhausting toil was distinctly harmful.

Most harmful of all was exposure. Even relatively short trips across the ice to aid other vessels might produce frostbite. Wooden walls — even with an extra lining of sailcloth — did not give much protection against cold so bitter that sometimes the mercury froze in the thermometer bulb, but they were much more effective than flimsy tents. When crews had to camp on the ice illness frequently resulted. After the *Thomas* was crushed on 13 December, it was several days before the re-allocation of her crew to other ships was complete. The death toll among these men was particularly high. On the *Dee* too there was a high mortality, perhaps related to the three day period on the ice in October when ice-movement threatened to destroy the vessel. Certainly, Ford Littlejohn, the surgeon, observed that they started to need attention almost immediately after going back on board.[67]

Roughly a month into their ice-imprisonment, the sun disappeared below the horizon, not to reappear for nine weeks. At first there was good strong moonlight, but by early December even that comfort was lost, though not for long. During its 28 day cycle there would be periods of ample light, at its best around 20 hours a day, but a cold light that did nothing for men's spirits.

Further north, the *Swan* of Hull lost the sun sooner and met it again later. She was in a worse situation — without any consorts. A valiant but foolish attempt by volunteers to seek aid from a Greenland settlement by walking 60 miles across sea ice led to the death of almost all.[68]

Little by little, health declined till the tell-tale symptoms of sore and bleeding gums peeling away from the teeth showed that scurvy had broken out. To the royal navy it had for two generations been a disease of the past with no likelihood "that ships or fleets would be crippled by this ancient curse of the sea".[69] The credit was due to the researches of James Lind, for 25 years Physician in charge of the Haslar Royal Naval Hospital at Portsmouth. In the French Revolutionary and Napoleonic Wars (1793-1815) the Admiralty had ordered the use of his anti-scorbutic remedies. As a result the navy had stayed fit for duty. Scurvy became a rarity whereas in previous wars it had on occasion destroyed the effectiveness of Britain's vaunted navy.

If it had taken the Admiralty forty years to accept the precepts of "A Treatise of the Scurvy", it took ordinary commerical interests much longer to learn. Scurvy badly afflicted civilian transports carrying supplies to R.N. blockading squadrons in the French Revolutionary War. In every year when whaling ships were caught by the ice and forced to winter in the Arctic, scurvy took a dreadful toll. Not till the Merchant Shipping Act (1854) were merchant ships required to carry anti-scorbutic supplies. Even then the requirement was not spelled out in detail. Hence their seamen had to wait a further 40 years to be relieved of that scourge.[70]

Scurvy is not an infectious disease; nor is it contagious; but due to a deficiency of Vitamin C. It produces horrifying effects on the human body, and ultimately death. Lind had never heard of vitamins. It would be a century after his death that they were identified, but he was well aware of the value of certain fruits, vegetables and herbs to prevent and cure the disease:—

> ". . . scurvy-grass, cresses, and other acrid alcalescent plants, are found highly anti-scorbutic, but it must likewise be remembered, that they are not perhaps altogether so efficacious as the acescent fruits; or at least become much more so by the additon of lemon juice, oranges, or a little sorrel".[71]

Lind knew that of these acid or 'acescent' fruits oranges were better than lemons and that both had much more efficacy than the limes that the navy later substituted for reasons of economy and which led to the recurrence of scurvy in the later nineteenth century.

If preparations were carefully made there was no need for scurvy to afflict an Arctic expedition even though cold and damp were noted by Lind as contributory factors.[72] Twice in the post-Napoleonic period Edward Parry made prolonged exploratory voyages into the Arctic. Only on his first expedition (1818-21) was there any scurvy and it was rapidly cured by liberal dosage with lemon juice. Parry expressed pride that his second venture (1821-3) saw no scurvy despite a 27 month stay in the Arctic. This was due to the large quantitities of anti-scorbutics carried, although Parry gave the credit largely to canned meat and vegetables. None the less he followed Lind's advice on both voyages by growing mustard and cress in his cabin.[73]

Whaling expeditions were not equipped with expensive novelties like tinned meats nor with lemon juice and preserved berries and vegetables. The owners supplied their vessels adequately with hard bread and salt meat, but also took advantage of the region to which they were sailing to hang quarters of freshly slaughtered animals at mast tops, dousing them daily with seawater until they froze. By the time the ships were held in autumnal and early winter ice the fresh food was exhausted. As the icebound men became less fit they also had to cope with an indigestable diet of salt beef and hard ship's biscuit.[74]

They had no source of Vitamin C. Eskimos in like situation would have sought a highly nutritious diet of raw meat and blubber, with raw seal's liver as an especial delicacy rich in necessary vitamins. The whalemen did not catch enough wild creatures. Even if they had, they would have been revolted by Eskimo ways.[75]

The journals spell out very clearly the disastrous effects on the crews of their ordeal. Meanwhile at home there was much press speculation and a good deal of concern about the missing ships. It was, however, left to an anonymous R.N. surgeon to express anger at the treatment of the whalemen.

> "The owners . . . could not say that such a prevalence of disease and such a mortality was not to be expected, when they had the melancholy example before them of the fatal effects of scurvy

> during the preceding seasons, notoriously caused by a want of nutritive food, and a total neglect of all means of cure. . . . although they were conscious that every ship was liable to pass the winter among the ice, did they make any change in the provisioning, so as to endeavour to preserve health in future?Did they use any precautions to prevent disease? Or did they send any better means to cure it when it did occur? No: These seem to have been matters of no moment with the gentlemen dealers in oil in the enlightened year 1837. . . Let us, therefore, appeal to the Legislature."[76]

The government too felt that it was irresponsible of the shipowners to create the same crisis situation all over again and therefore hastened to deny any involvement. "The Lords Commissioners of the Admiralty" did, however, offer the consolation that they "cannot believe that the owners . . . can have sent them to sea without an adequate supply of provisions. . . . The owners of the ships now detained in the ice, and other persons connected with the trade, should lose no time in sending the means of succour, similar to that which was afforded last year to Stromness."[77] As a sop a surgeon's help was offered. After all, there were plenty on half pay. Otherwise, self help was the thing. "It is no part of the duty of the Admiralty to establish an Hospital at Stromness, nor have they funds properly applicable to the purpose."[78]

Gradually government attitude softened — as the Shipping Gazette's hopes of action were widely reported by the press;[79] as former premier Sir Robert Peel presented a petition from Dundee in the House of Commons; as supporters like A. Bannerman, M.P. for Aberdeen burgh, gently prodded them; and others like Cllr. Bisset vigorously declaimed against inertia.

> "The families of 300 men were at present in a state of extreme poverty and distress — and Government had thought proper to insinuate that it was the owners of the vessels that should be at the expense of sending out aid to the sailors . . .

Hull whalers *Jane, Viewforth* and *Middleton* beset in 1835.

Adam Flett, a survivor of the *Dee*.

'The kirk on the hill' — St Michael's, Harray.

he graves of Adam Flett and James Tulloch in St Michael's kirkyard, Harray.

Handest, Harray.

Nistaben, Harray.

> the owners who suffered so much in the business, and who had to give the monthly payments to the families of the seamen. . . . The relief of the seamen was a national object, and the nation ought to pay the expense."[80]

After the previous year's experience it was plainly useless to send a ship on search duty. The best prospect lay with ordinary traffic. Bounties were, therefore, promised to the first five whaleships provided they sailed a lot earlier than usual, bearing relief supplies for the missing vessels. They would also be handsomely reimbursed for all aid actually supplied.[81] There was a good response, but bad weather played a role in keeping vessels in Stromness till late February. The *Traveller* of Peterhead had arrived on the 7th. and was "completing her crew at 40s. per month, and 1s. 3d. per Ton. — The *Traveller* is the first vessel that has been dispatched in search of the missing Whalers, and carries out extra hands, stores, &c." She sailed nearly a fortnight later, accompanied by the *Princess Charlotte* and the *Horn* of Dundee, each of which had taken 10 men in Longhope.[82] Some of the beleaguered ships benefited greatly from this aid. For instance, the *Lord Gambier* supplied the *Norfolk* with gratifying results for "several who were previously confined to bed were, by the use of fresh diet, soon enabled to come on deck and resume their duties."[83] The *Grenville Bay* met with no fewer than seven helpers. In consequence, not only did she gain ample fresh foods of which the most valuable would have been potatoes and cabbage, but also sailed on with six new, active hands.[84]

When the ships limped in to Stromness it was to meet with another indication of a softening official attitude. A hospital had again been opened. The evidence for 1837 does not establish either building or surgeon. It is likely that Mrs Humphrey's house would still be empty and available for hire; Hamilton was in all probability still in Stromness, having ended his unhappy naval career. Aged 26 and already medically qualified he had entered the navy in 1826 as an assistant surgeon. Such a post gave low pay, poor status, no opportunity for study and, in peacetime, scant chance of promotion. He shared living quarters in the dingy cockpit with midshipmen who might be only 12 or 14. In consequence, wrote one pamphleteer, "they prefer the permanent establishment of a village surgeon, in some remote and miserable place, to the service of the State."[85] The Admiralty may have preferred to send one of their current staff, but it is possible that they would have thought Hamilton's recent experience of scurvy too valuable to be wasted because of pique at his frequent pleas for lengthy periods of half-pay leave.[86]

The press had been starved of news. As the ships arrived they used every scrap — from ship agent, Margaret Login's letters announcing the arrival of the *Dee,* to the lengthy letter from the *Dee's* surgeon, Ford Littlejohn to Rev. Dr. Paterson of Montrose. The papers shamelessly copied each other so that the country was quickly aware of the happenings of the winter — Littlejohn's letter, and comments that he had apparently made in Stromness, being the major source. Conditions had become appalling. "Two or three were lying together in one blanket, covered with ice, and the blanket underneath literally a mass of vermin. The dying were often lying in the same bed with the dead for days together; and when obliged to consign the latter to the deep, the bodies had to be hoisted up the 'tween decks with a tackle and thrown overboard."[87]

By May the main drama was past; there was only periodic mention of the still missing ships, once plumbing the depths of banality with a report of Stromness rumour:— "There is no information of the *Advice* or *Swan,* whalers, and the general opinion is, that there never will be any account of them."[88] Not till summer were the last acts of the tragedy worked out. In June, after yawing helplessly about the western ocean the *Advice* arrived in Sligo with a bare handful of survivors. Weeks later the *Swan* seemed raised from the dead when a memorial service in July was interrupted by the news that she was close to the mouth of the Humber.[89] Probably this was the worst Arctic whaling tragedy; unfortunately, it was not to be the last.

Postscript — The Adam Flett Story

Often, in episodes of high drama, the ordinary man has his moment of attention. He

"... struts and frets his hour upon the stage
And then is heard no more."

So it was for the most part in this case, but we are fortunate to be able to follow more of the life of one Orcadian survivor.

Adam Flett (1812-1894), sixth son and eighth child of James and Margaret Flett of Linneth in Harray, had little expectation of much inheritance. He would have to make his own way in the world. Thus he left his land-locked parish to work in the slate quarry by the Black Craig outside Stromness burgh. At the age of 17 he left that employment for the first of several whaling voyages, the last being the harrowing expedition aboard the *Dee* in 1836. When she returned he was one of "only three men ... able to go aloft or carry a bucket of water."[90]

How was it that Adam was still reasonably fit when all around him was death and despair? He himself credited his good health to running each day on the ice till he was warm, which was a vastly better way of spending time than sitting bewailing the cold and hunger. The effects of moping were disastrous. On that, sea captains and medical men were agreed:—

> "Whatever discouraged our people ... never failed to add new vigour to the distemper [scurvy]: for it usually killed those who were in the last stages of it, and confined those to their hammocks who were before capable of some kind of duty."[91]

The impression of his descendants is that Adam Flett accepted life as it came, neither over-excited by its triumphs nor greatly upset by its troubles. Such evenness of temperament would be invaluable in a situation where one could do nothing but wait. Stress and anxiety burn up Vitamin C at a fast rate.[92]

Long afterwards Adam was wont to comment to his sons about the sign that convinced him that escape to the

(Y) Certificate of Character.

Name of Seaman.	Aaron Flett
No. of Register Ticket.	134.065
Name of Ship, and Port of Registry.	Harmony London
Name of Master.	J. White
Description of Voyage.	Labrador
Date of Commencement of Voyage.	5 June } 1881
Date of Termination of Voyage.	6 November } 1881
Character for Ability in Seamanship.	VG
Character for Conduct.	VG

I CERTIFY the above to be a True Copy of so much of the Report of Character, made by the said Master on the termination of the said Voyage, as concerns the said Seaman.

Dated at 49 Minories this 14 day of November 18[illegible]

Signed [illegible] *Shipping Master.*

NOTE.—Any person who fraudulently forges or alters a Certificate of Character, or makes use of one which does not belong to him, may either be prosecuted for a Misdemeanor, or be summarily punished before a Magistrate by a penalty of £50, or by imprisonment with hard labour for three months.

By Authority: Printed for the Board of Trade, by James Truscott, Nelson Square, London.

open sea was close. They had drifted south, held tight in the embrace of a great ice sheet, when they noticed the lamp begin to swing — an indication that, after five months, moving water was starting to influence their destiny.

There followed the long depressing struggle to sail a grossly undermanned ship back across the Atlantic, made worse by the fact that no-one who knew anything of navigation was fit to leave his bunk. Near the Hebrides, when utterly exhausted, they met in quick succession displays both of human cruelty and of human compassion. One ship ignored their distress signals and sailed away. A fishing boat accepted salt beef from them, but gave in exchange none of the fresh fish they craved. Probably a language difficulty between Gaelic-speaking Lewismen and broad Orcadian and Aberdonian whalers occasioned that, but, for some reason, Adam got it into his head that the culprits were fellow Orcadians from the island of Hoy. Ever afterwards he had not a kind word for Hoymen. Soon they met the *Washington,* outward bound across the Atlantic. Captain Barnet took them in tow and delivered them to the "charge of Lloyds' agent and branch pilots"[93] who took the ship into Stromness harbour.

Adam Flett remained at sea, but never again on a whaler. When he did return to Arctic waters it was as a seaman aboard the *Harmony* carrying missionaries and mission supplies. He must have been of that canny breed of Orcadians who saved carefully for the day when he could have his own piece of land for, while he was at sea, the small farm of Handest was bought on his behalf by a brother. Adam settled down there, marrying his near neighbour, Anne Flett of Nistaben.[94] After all his adventurous career on the face of the ocean, he had returned to his native parish of Harray, the only one in Orkney that has no sea coast.

ACCOUNT I

The Dee

Adam Flett Hauder Harray one of the surraiving sea men on bord the Dee tak care of thes Book

A

NARRATIVE

OF THE

SUFFERINGS OF THE CREW

OF

THE DEE,

WHILE BESET IN THE ICE AT DAVIS' STRAITS,
DURING THE WINTER OF 1836;

WITH

OTHER INTERESTING AND IMPORTANT PARTICULARS, DRAWN
UP FROM NOTES, TAKEN AT THE TIME, BY ONE OF
THE SEAMEN ON BOARD.

ABERDEEN:
PUBLISHED BY GEO. CLARK & SON,
No. 15, BROAD STREET,
FOR THE BENEFIT OF DAVID GIBB, SEAMAN ON BOARD OF
THE DEE.

MDCCCXXXVII.

INTRODUCTION.

Loss of life, under ordinary circumstances, is simply remembered with common feelings of regret; but the disastrous consequences of this unfortunate voyage are of too melancholy a nature to be easily forgotten; and, while the relatives of the deceased may derive a painful satisfaction from the perusal of the following pages, public sympathy in their behalf may also be stimulated. Moreover, should it appear evident that the sufferings and privations of the seamen were in any way occasioned by causes which might have been in some measure avoided, the narrative will be useful to others who may possibly be placed in similar circumstances.

TO THE READER.

SINCE the present publication was announced in the *Aberdeen Herald* newspaper, so many inquiries have been made regarding its nature and authenticity, that there is reason to believe the public will take a deep interest in its contents. It is, therefore, the more necessary that a few words of explanation be subjoined.

David Gibb, the individual for whose benefit the narrative is published, was one of the seamen on board the Dee. It was his privilege to possess, during the whole of the season, much better health than very many of his comrades; and, having received an ordinary education, he was the better able to record faithfully the trying circumstances of the unfortunate voyage. It is but right to state that the surgeon, Mr. Littlejohn, gave him much assistance. Since the arrival of the Dee, Gibb has been requested to put his manuscripts into the hands of some one more experienced in literary matters. This was done; and, if the narrator has been enabled to discharge the duty with satisfaction to the public, he will be fully gratified.

J. H. W.

12, CONSTITUTION STREET, ABERDEEN,
25th May, 1837.

NARRATIVE, &c.

The Dee sailed from Aberdeen on the 2d of April, 1836, having on board thirty-three officers and men, and the usual quantity of provisions. On the 7th, engaged sixteen additional hands at Stromness, proceeded to Davis' Straits on the 9th, and on the 15th of May made the ice. Next day, coiled the lines; and soon after reached up the east side, but saw no appearance of fish. For two or three weeks, the weather was rather unsteady; but the progress upwards was chiefly obstructed by loose ice and icebergs; and so numerous and dangerous were the bergs on the reef-coal, in lat. 66°, that serious fears were entertained for the safety of the vessel. At times, it was almost resolved to return and proceed to the west ice; but, although several vessels determined on this course, Captain Gamblin seemed exceedingly anxious to get the Dee north. With this view, she was anchored by an iceberg at Hair Island; and on the earliest opportunity, advanced as far as Fair Island Point. Here she lay a fortnight; and the ice having then opened, North-east Bay was gained without farther interruption. Here again the ice closed; but, contrary to expectation, opened in a few days, and allowed the Dee to proceed as far as Frow Islands. There was very little ice in this quarter, and the prospect west being rather tempting, an effort was made to get through; but, after considerable progress, the difficulties were found to be too formidable; and, in consequence, the course was changed to about E.NE., under the expectation that the North-water would, in this way, be reached. This course was more fortunate. After about two days' sailing, she made the fresh ice in company with several other vessels — on reaching which, the Dee was moored to a floe. In the course of a few days, the ice opened; the Swan of Hull went first forward, observed the North-water, and made signals accordingly. The Dee, in company with ten more, took the ice, and got through without danger. The course was now shaped for Pond's Bay; and, with the exception of one heavy patch of ice, no obstruction worth noticing happened during the passage.

Pond's Bay was reached on the 12th of August; and, on the same day, Captain Gamblin discovered fish by the land ice.[1] All hands were in consequence called on deck, boats manned, and every necessary preparation made for immediate fishing. On the 13th, the first fish was struck; and, the weather being fine, she was easily secured. The whales were rather plentiful in the Bay; and, during the remainder of the month, other three were killed, and three dead ones picked up. Captain Gamblin, however, as the season was getting through, thought it would be advantageous to try a little farther south. This was attempted; but, after getting down as far as 71°, the ice was found to be very heavy, and no fish could be seen. Pond's Bay was again reached; but, on arrival there, it was found that the fish had left. By this time, a good many vessels were in the Bay, among which was the Friendship of Dundee[2] with fifteen fish. This vessel had never left the Bay.

Believing that no more fish would be killed here, Capt. Gamblin thought it prudent to return homewards. The wind was now SE.; weather moderate, and, after proceeding a short way, the Dee fell in with the Grenville Bay, the captain of which stated that he had been as far east as to see the Duck Islands, but found the bay ice so heavy, and making so strong, that a passage that way was deemed impracticable.

It was now the 13th of September, and, fearing the worst, all hands were called on deck, and a proposal made to go on short allowance. To this every one agreed; and the quantity of provisions being stated, the mess was limited to four pounds of bread, per week, for each man, and a corresponding reduction in beef, meal, barley, &c. Captain Gamblin determined on trying the north passage again, and succeeded in getting as far as 75°, in company with the Grenville Bay and Norfolk of Berwick. The east land was here seen, Cape Melville[3] distant only thirty miles; but the bay ice was making too strong to admit of farther progress. All hands were again called on deck, and a consultation held as to the best means of effecting a passage. Captain Gamblin stated that, on considering the matter along with the captains of the Grenville Bay and Norfolk, they had thought proper to advise another attempt south, keeping a clear side as long as possible. To this proposal every man agreed, and accordingly the Dee bore away on the 20th; wind NW.; easy weather. On the 23d, got into lat. 71° — the bay ice very strong. Early in the morning, fell in with the Thomas and Advice of Dundee, and were assured that they had also been as far north as 75°, but saw no prospect of getting through. They had seen a berg where eight vessels had been made fast, as shown by the marks of ice-anchor holes. Towards night, the bay ice was making so strong that Captain Gamblin ordered the Dee to be made fast to some sconce pieces; the other vessels did the same. In the morning, a good deal of water was seen, bearing about SE. and S.SW.; but, after proceeding in that course for three days, it was found impossible to proceed farther.

Here again was a "black look out;" and the prospect of a passage being now almost hopeless, it was thought right to submit to a farther reduction of allowance. Accordingly, on the 27th, the mess was fixed at three pounds of bread per week, with a proportionable reduction in the other provisions. From the 27th to the 30th, the Dee remained in the same latitude; but, the wind prevailing from the east, it was again arranged, after some deliberation, that all the five vessels should proceed northward; the Advice and Thomas took the lead, and soon outreached the others. On the 1st of Oct. the two Dundee vessels were again seen. The weather at this time was very bad, the winds being generally from E.NE. with snow; the ice, too, was so strong, and the swell so heavy, that none of the vessels could get farther north. All hands were again called on deck — the painful circumstances of the case fully considered; but the season being now so far gone, it seemed a matter of indifference as to what course should be pursued. On the 2d, Captain Gamblin went on board the Grenville Bay, had a consultation with Captain

Taylor, hoisted signals for a conference with the other vessels, but by some means or other they were not perceived. Captain Gamblin then returned to the Dee, ordered the vessel's head to be laid south, with a view to proceed to as safe a wintering station as possible; but the wind failing, she could advance no farther. During the next three days the wind freshened, and the ice opened up in several places. On the 8th, took the sun's altitude, and found the ship to be in lat. 73° 12′. The other vessels were also fast. On the 10th, got another observation, and found that the drift had been two and a half miles south. The ice was now so strong that the men could travel from either vessel without danger; but the wind having taken a hold in the N. and NE., the swell outside caused frequent interruptions, and more particularly in the immediate locality of the Dee.

From this date, the peculiar sufferings of the crew may be set down. There was not any farther reduction of allowance; but the fires having been extinguished from 8, P.M. till half-past 5, A.M., the beds got exceedingly damp and uncomfortable; indeed, so much so, that there was a constant dropping in every one of them.

The chief points of consideration now, were the health of the crew, and how to keep every one in as active a state as possible. For this end, a variety of work was performed, which, in other circumstances, would have been perfectly useless. There were some things done, however, which, on account of the peculiar associations they called up, are worthy of particular notice. Such, for instance, as the sending down of the top gallant yards, unbending several of the sails, and unshipping the rudder. Seamen who have been accustomed to these duties know well that they are generally the harbingers to a safe and comfortable wintering; and, under such a prospect, it is pleasing to discharge them. But, alas! how different were the feelings of the seamen on board the Dee on this occasion! They had no pleasing hope of spending a happy winter with those who were near and dear to them — no joyful prospects of returning spring — no home — no comfort — no delightful enjoyment to anticipate at their own fireside. Instead of this, while the sails were unbending, and the halliards unwillingly yielding to the unreeving hand, the tears of sorrow and regret were mingled with the sighs of forlorn and almost desperate hope.

But the time was very soon to be otherwise occupied. The ice, as has been already mentioned, being so loose about the Dee, a fatal squeeze was hourly anticipated. This state of anxiety led to almost constant watchfulness, and the frost having become very hard, exposure was the more to be dreaded. On the 15th, the sun was again taken, and the latitude ascertained to be 72° 58′. On the 16th, it was 72° 50′ — wind strong at NE., and large icebergs floating past. About this time the ice began to press hard, and was particularly severe, end on. During the night of the 16th, the vessel was hung by the quarter, the ice squeezing all along as high as the guard boards. At daylight, the captain called all hands on deck, and ordered every one to lose no time in getting up the provisions. At 8, P.M. the wind fell off, the ship still hung up by the quarter. The ice, however, was quiet, and the prospects being rather more satisfactory as night fell down, several of the crew returned to take what rest they could get. At 11, P.M. there was another dreadful crash, but it passed over with less fearful consequences than were at first anticipated. On the 18th, the ice gave way in several places, and opened up so far that a warp had to be got out to secure the Dee. All this time the other vessels lay undisturbed. Fresh water being much wanted, a boat was dragged across the ice to the nearest berg, distant three miles, and such duty was repeated frequently, and had a very baneful effect on the health of those who were thus employed. On the 20th, the ice closed again with some sharp nips. To strengthen the ship, ten extra beams were put in aft, and the casks stowed in such a manner as to assist also in resisting the pressure. The strengthening was very seasonable, for the very next morning a very dangerous crash was felt; it ranged both fore and aft, and, for the time, led every one to think that the vessel was gone. It passed away, but only to give place to another in less than half an hour. This was a still more dreadful squeeze, and, under the impression that all was over, bags, chests, and every thing that would lift, were in one moment on the ice.

The dreadful sufferings which were this night experienced cannot be described. To enable the reader to form some idea of the situation of the men during this period, let him imagine one field of ice, of almost immeasurable extent, studded here and there with icebergs towering to the clouds. In a small spot is fixed the Dee, now reeling to this side and now to that, and every alternate roll attended with a crash, the sound of which more resembled the convulsive groans of an opening earthquake than the natural dashing of displaced water. On both sides, the crew are ranged at sufficient distance to avoid danger from the falling of the masts, and without any shelter, or fire, or other protection from the freezing elements of nature, now severe in the extreme; and this, too, during the solitary hours of night, and without that comforting hope which is engendered by the prospect of an early dawn.

Such, then, were the circumstances in which the crew of the Dee were placed on the 20th of October last, and human nature under these could not fail to quail before them. Contrary, however, to all expectation, the Dee was not injured, but, fearing another crash, the crew resolved to remain on the ice for a day or two more, now and then going aboard as the vessel eased. On the 22d, an observation was taken, and the latitude set down at 72° 11′. At ten at night, the ice broke up and drove, till about two next morning. Some dreadful crashes were then felt, and under the belief that no security was at hand, as much additional provisions as could be at all laid hold of were taken on the ice. While thus engaged, the most threatening squeeze yet experienced racked the Dee fore and aft. For safety, every man fled as he best could, and again under the dread of almost hopeless escape from the tottering masts. By and by the ice fell quiet, and as much of the provisions, &c. as possible, were again put on board. The pumps were tried, but no water was in the vessel. On the morning of the 23d, a good many lanes of water broke out; and here again the comparative comfort which had been for a few days enjoyed was disturbed; for it is invariably the case that, when these lanes make their appearance, the ice becomes very unsafe. Under the impression that another squeeze was pending, all hands were ordered to go on the ice with saws, and cut it up in as small pieces as possible. This was done, and some pieces being parbuckled and others sunk, the vessel was cleared a few feet. The Dee was farther eased at this time by the unexpected opening of a floe on the starboard side, and as soon after as possible she was hove a length in head. For a few hours, the greater part of the crew went to bed; but, at three in the morning of the 24th, she was squeezed again, by the one floe overleapping the other, and raising the vessel some two or three feet. Next morning the ice took off a little, and, as it appeared to be thinner right a stern, the Dee was hove down about a hundred yards. What of the boats, chests, bedding, &c. as were not got on board at the former berth had to be dragged along on the ice to where she was now moored, but the difficulties in this duty can only be known by those who have had them to encounter. Everything being again on board, and the ice getting more settled, hopes were entertained that the vessel would lie in safety until the spring.

The 25th was a very disagreeable day, wind NE., blowing very strong. The most painful feelings of this day arose from the fear of the wind prevailing from that quarter, the North-east Water being very near, and the consequent irruption of the ice inevitable. Captain Gamblin here resolved to make every exertion to cut a dock for the Dee. With this view, several men from the Grenville Bay lent their willing assistance, and with the aid of ice-saws and other means, a dock was cut out. This operation, however, cannot be passed by without noticing its nature and effects on the men. First of all, a triangle had to be raised on the ice, and a block fixed in the junction at the top. Through this block a rope is reeved, to which an ice-saw is suspended, and from the other end bell or branch ropes, to allow five or six men to work the saws, and which is done in the same way piles are driven for laying foundations of quays or other buildings. The saw being raised as high as may be necessary it is permitted to fall, and its own weight cuts downwards. When a piece of ice, large enough to be sunk, is cut out, as many of the men as can stand on the edge next the solid floe, weigh it down, and when the surface edge underlaps the floe, it is pulled in and kept down by the pressure of the heavier ice above. The great danger in this operation consists in the exposure of the men to cold by standing in the water, for it not unfrequently happens that the ice must be sunk so far as to permit the water to enter in by the top of the men's boots. Several of the crew of the Dee felt the effects of this, in having their feet frosted, and their limbs rendered almost helpless. But to return, the frost was becoming every day more severe and so powerful, that the 'tween decks were all white as snow, and near where the fire had been through the day, icicles were found during night.

On the morning of the 26th, a solitary bear was seen; but just at the time when the crew were expecting to be benefited by such fresh provisions he made off, thus blighting all anxious and pleasurable anticipation, and inducing the awful conviction that every thing was conspiring to accelerate the fatal end of every one on board. The next three days passed away much more comfortably than could have been expected — the weather having moderated and the ice kept firm. During this period, the men were chiefly employed in carrying ice from the bergs to dissolve into water; but, unfortunately, while one of the boats were being dragged across, the keel broke, and thus created additional labour and fatigue. Thus far the men were in good health, and under the prospect of being able to repeat the visits to the icebergs; a sledge was made, but, alas! never used for that purpose, as will be seen by and by.

On the 29th and 30th, there was a change of weather for the worse; but the ice did not rise so much as was expected. On the latter day, three bears were seen and fired at, but without effect. This was trying, as fresh provisions were much wanted; but there was one pleasing circumstance in connexion with this disappointment which is worth notice; two men having occasion to be on the ice perceived the bears, but fortunately made their escape in time to save their lives. It was but a respite, however, for the poor fellows, soon after fell victims to the unrelenting hand of death, though laid on with milder severity.

November opened under rather more comfortable prospects than its predecessor exhibited. There was now no hope whatever of relief before the spring; and the ice being much more firm, the men had less harassing duties to discharge. But even with these mitigating circumstances, what can we say of comfort in a latitude of 72° 50′ — forty-nine men on a miserable pittance of provisions — beds and blankets beginning to freeze with ice, and little or no fire to dispel the cold. On the 2d of this month, a sad reverse was experienced, in the dock giving way in several places, and the ice again threatening to crush the vessel. Captain Gamblin, however, resolved on cutting a new dock; and, having again procured assistance from the Grenville Bay and also from the Norfolk, forthwith set to work. The details of this labour being already stated, repetition is unnecessary; suffice it to say that the dock was prepared, and the Dee, with much difficulty, got moored therein. On the 3d, 4th, and 5th, the weather became exceedingly boisterous, and more snow fell than during any corresponding period since the vessels were beset; and what made the situation of the men more painful, the supply of coals was now nearly exhausted. In the meantime, one of the boats was broken up, and soon after another shared the same fate. On the 6th, the dock gave indications of rending; and, to provide against so fearful and so frequently fatal an incident, all hands were ordered to cut the ice in end as far as thirty fathoms. This was done; but the frost became so intense that the first length closed before the vessel could be hove in. This day, another observation was taken, and the latitude found to be 72° 23′. In the afternoon, the wind veered from SW. to SE., and towards night the ice opened in all directions. Serious fears were again entertained for the safety of the vessel; and, at night, so sharp were the nips, that more than once all hands were prepared with the bags again to go on the ice. At 11 o'clock, destruction seemed inevitable; but, soon after lowering the boat and leaving the vessel, the ice immediately closed, and thus permitted the men again to go on board. The provisions were farther reduced on the 7th, the heaviest reductions falling on the beef and barley. On the 8th and 9th, the ice was very unsettled, and little or no rest could be got for fear of a squeeze. Several of the men were severely frostbit in the face at this time, and relief was only got by rubbing with snow the parts affected. In the afternoon, two bears were seen, but like the others made their escape. The morning of the 10th looked well; and Captain Gamblin thinking a safer dock might be cut, assistance as formerly was procured from the other vessels. The dock being finished, the Dee was hove in, and the ice to all appearance was more secure than it ever appeared before. On the 11th, the observation showed 72° 15′; the wind moderate and sky clear. In the forenoon, the ice gave way alongside the Norfolk, and a watch from the Dee went over to assist in cutting a new dock. On this occasion, several of the men were again frostbitten. In the evening, two foxes were seen near the Dee, but they also made their escape.

By this time, however, several of the men had managed to cook the tails of the whales so as to form a tolerably agreeable meal. But the fact is, very few were now fastidious in their taste, and it might have been well had every one endeavoured to accustom himself to the provision in question; those who did so felt themselves much benefited. For the advantage of others who may happen to be placed in the like distressing situation, it may be well to give a brief outline of the manner in which the tail was cooked. The detail, too, will be interesting to the general reader. In the first place, it was cut up into pieces about four inches square, then laid out on the ice to bleach, or be purified by the frost, where it lay for four days; next, parboiled, then allowed to settle on the top; the top was then scummed off, the pieces lifted out and laid past till they were wanted; and, when about to be used, were fried in the pan with a little fat. Delicacy, it will thus be seen, was here out of the question; and, however exceptionable the doctrine of the "end justifying the means" may be, it will be obvious to every one that, in the present case, it was justifiable.

On the 12th, the sun was almost gone, and the weather excessively cold. Captain Gamblin, in the forenoon, gave each man a yard of canvas to make into snow-boots with wooden soles, which proved of great benefit to the men. Had it not been for this, many more would have had their feet frostbitten. To allow flannel to be rolled round the feet, the boots were made very large, and the wooden soles were remarkably well adapted for keeping out the frost.

The 13th being Sunday, Mr. Littlejohn,[4] the surgeon, read two sermons and prayers; and the value of religion seemed more to be felt at this moment than ever it had been before. The crew had been very anxious, for some time, to have religious exercise on board; but at first they were rather diffident in requesting Mr. Littlejohn to take the lead. A deputation, however, waited on him, expressed their anxiety on this head, and it says much for that young gentleman that he readily and most cheerfully complied with the request; and not only did he lead the exercise on Sundays, but also on other days, more particularly on Wednesdays and Fridays. Here we must pause for a moment, and allow the reader to reflect on the solemnities of such a scene. Think, then, of forty-nine fellow-beings assembled together in one common bond of Christian worship; but no gilded temple shone around the worshippers — no taught minister of salvation declaimed from the pulpit — no eloquent oration, the production of calmful and uninterrupted meditation, fell on the ears of the assembly — nor any church-going bell tolled the hour of dismissal. But what is worth more than all these, there was but one sentiment, one feeling, one motive, and one object in view; and this one sentiment and feeling and motive and object were the offspring of a sincere conviction that the realities of eternity were soon about to burst on them. It were degrading to human nature to assume for a moment that *now* everlasting peace was a consummation to be viewed with apathy or indifference; and, therefore, while we willingly assume that, in this case, every soul was sincere, let us hope that those who have gone to give their account have done so "with joy and not with grief."

On the 15th, the sun did not make his appearance. This was a sorrowful day. Each one surveyed the heavens, alas! with anxious thought, and but too sensibly observed a "blank in nature." It was a change which none on board had ever experienced; and what rendered the contemplation more painful, was the very distant hope of ever seeing his brightness again. Different circumstances combined to mature this anticipation. Those more particularly worthy of notice were the immediate intensity of frost, and the threatening appearance of the ice. With regard to the first, it was so severe that few could sit for an hour on end and rise with anything like freedom; and, with respect to the latter, the North-east Water still told with fearful effect on the ice by which the Dee was surrounded. From the 16th to the 30th, nothing calling for particular notice took place. On the latter day, Captain Taylor of the Grenville Bay took an observation, and set down the latitude of 71° 57′.

December opened with gloomy propsects; the effects of so much exposure and harassing duties being now severely felt. The first indications of disease were coughs, swelled limbs, and general debility; and it is worthy of particular notice, that the flesh became discoloured with small red spots, attended with sharp pains and stiffness of the limbs. Depression of spirit invariably followed when these symptoms of disease were perceived, and even with the opportunity of active and healthful exercise there was little or no inclination to profit therefrom. On the 3d, shot two foxes, but "what were they among so many?" Those, however, who did partake of them, expressed their confidence of having been much refreshed. On the 5th, another fox was killed, and the appearance of several more induced the hope that a supply of fresh provisions might yet be at hand. On the 8th, another observation was taken, lat. 71° 12′. At this time, the wind was easy, the sky clear, and the northern lights very brilliant; but were never heard so distinctly as on the Western Ocean.

The Thomas of Dundee was listed over on the 12th by a heavy pressure of ice, and so far did she heel, that the crew were obliged to creep on their hands and knees over the deck. This situation, distressing as it would have been at any time, was now much more so, the frost being so severe and the daylight almost gone.

On the morning of the 13th, a melancholy spectacle presented itself. But a few hours before, the Thomas, though heeled, was not damaged; but now she lay a total wreck. The masts were not yet gone by the board, but the vessel was past recovery. About mid-day, twenty-two men went over from the Dee to the Advice, which lay between the Dee and the Thomas; but being unable to proceed father that night, resolved to abide there until morning. The crew of the Advice at this time were pretty healthy; but the provisions were so short, that the men from the Dee could only get one half-pound of meal each man for supper, and the same next morning. At eight o'clock in the morning of the 14th, about sixty men from the Dee and Advice went over to the wreck of the Thomas, with a view to save as much of the provisions as possible, and render assistance to the men in saving their clothes, &c. Three days were spent in this hazardous work, and when all were collected, an equitable distribution among the whole of the vessels took place, both of men and provisions. It is right to notice that two of the crew of the Thomas died on the ice the night she was wrecked. These were the first deaths; and while the surviving seamen took a last farewell of their departed comrades, they could not help thinking that a similar fate, though perhaps under different circumstances, awaited themselves. While engaged in carrying the provisions from the Thomas, much injury was sustained through the cold. The ice, too, over which they had to pass, was exceedingly unequal, and, in different places, was divided by lanes of water. In many cases, the men sank down into the water, and thus laid the seeds of that disease which soon after proved so fatal. The supply of provisions which the proportion of the Thomas's crew retained was much about the same as the mess for those of the Dee. The chief point of regret in regard to the loss of the Thomas was the distance; because that precluded the possibility of breaking her up for firewood.

A lunar observation was taken by Captain Taylor on the 16th, and the latitude set down at 70° 29′; the weather very moderate, and wind north. On the 17th, the latitude was 70° 12′; the sky clear, and a good deal of daylight. About this time, Captain Gamblin suggested a farther reduction of allowance, but the crew would not on any account submit.

The death-monster scurvy now began to harass the greater part of the men. The disease first appeared in the mouth, and was known by a swelling of the gum and deadening pain, which increased to an excruciating feeling when anything touched the parts. When beef was used, the suffering was dreadful, and the salt and cold frequently caused the blood to flow. On the 18th, twenty-one men were ill of scurvy, some of them suffering most severely. To add to the misery of all on board, the ice again gave way and threatened to squeeze every one of the vessels. The Advice seemed to suffer most; and, indeed, so dangerous was her situation at one time, that all hands were ready to spring on the ice, where the bags and a good deal of her provisions were already placed. The Grenville Bay was in great danger on the 18th, the ice having opened and squeezed as far up as the stern windows. Captain Taylor had almost despaired of her safety, as was indicated by a light on the bow, the concerted signal of danger. A great deal of snow fell on the 23d. The 24th is remembered chiefly on account of another reduction of the allowance. The bread was not reduced, but the pork and beef were. The quantity weighed out was half a pound of pork each man per day; and, alternately, with three-quarters of a pound of beef. From this date to the end of the month, several of the crew were confined to bed by scurvy, and their situation there was anything but comfortable. By an observation on the 30th, the latitude was ascertained to be 70° 13′; wind, E.SE. blowing fresh.

The 1st January, 1837, was a day of sorrowful remembrance. Fain would the mind have banished the recollection of a new-year's day at home; but every effort seemed fruitless. Memory conjured up with avidity every pleasurable enjoyment, and seemed, as it were, to revel in delight when the conflicting remembrance of past and present experience clashed together; and, O how dismal the contrast! On that day twelvemonths, almost each one could boast of a home, and not a few enjoyed the endearments of social bliss. Now, the husband had no smiling partner to comfort him in the intricacies of life, to soothe him in his sorrows, to enliven his solitude with the balm of conversation, and render his home the soft green on which his mind would love to repose; and now, too, he had no little ones prattling at his knee, lisping their affectionate attachment in words of "softest tenderness," and gladdening his care-worn heart with their smiles of innocence and filial obedience, And again, the youth in "beauty's pride" had no fond one near on whom his best earthly hopes centred, and whose reciprocal and faithful and tenderly-devoted attachment was the pleasure of his life and the ardent delight of his hopes and prospects. Or it may have been, and indeed it was, that the dutiful son was now deprived of another marked period when the most valued testimony of his sincere regard would have been tendered to a doating father or mother. Instead of this, there was nothing but blackness and darkness around. Disease had fixed his fell grasp on his intended victims; the writhings of his torment pierced the very soul, and despairing hope almost changed the best feelings of nature into the fiendish passion of a maddened brain. And farther, the elements of nature combined together, as it were, in order to complete their misery. Aloft, the tempest waged a dismal warfare with the yielding and tottering masts, and the wind wildly screamed through the frozen shrouds as if the death whistle had already sought to disturb the peace of the mariners below. Nor was this all, the crashing of the contending bergs as they rushed against each other and burst asunder, seemed to shake the very ocean, and spoke like the yielding forest in the mastery of the gale. In short, in every point of view this was a memorable day. and although many would have held it so, had they been spared, it is still one of deep interest to the surviving few.

The weather on the 2d, 3rd, and 4th of this month was very moderate; but the frost still exceedingly hard. Scurvy was daily making rapid strides among the men, and seemed to threaten universal destruction. Mr. Littlejohn the surgeon applied the most effective medicines he had, and continued his unwearied exertions with a degree of anxiety and care which does him the highest honour. But the disease could only have been cured by fresh provisions; and, alas! the prospect of such a remedy was distant indeed.

It occurred to the men, on the 5th, that the quantity of provisions on board would allow some additional allowance, and, with this view, a deputation waited on the captain to make the request; but Captain Gamblin thought it would be premature to interfere with the present mess, and, therefore, declined to grant the request. He mentioned, at the same time, that he was quite satisfied his men had the power to compel him to comply with their wishes, but he would hope that they knew themselves better than to attempt such a line of conduct. He was right in his conjecture; nothing was farther from the intention of the seamen than to disobey their commander's instructions, and, therefore, they would rather starve than take by force what was not given of pleasure. These sentiments being expressed to Captain Gamblin, he reconsidered their application, and ventured to add a little additional flour.

The brilliancy of the sky on the morning of the 6th, gave hope that the sun was soon about to make his appearance, and as the little daylight there was increased in brightness, the watch were gratified with a view of a large sheet of water on the starboard quarter, distant only a few miles. Such a scene earlier in the season would have been viewed with dismay; but now the possibility of relief was, to say the least of it, more rational. These anticipations, however, were too good to be realized and too pleasing to be gratified.

On the watch being called on the morning of the 7th, a large proportion of the men were unable to leave their beds. This was a melancholy state of things; and what now made the case still more painful was the fact that the beds were in a most deplorable state of cold and also of vermin; but more of this by and by. On the morning of the 10th, Capt. Gamblin ascertained the latitude to be 70° 18′. The weather was very moderate, but a good deal of snow fell.

Hitherto disease had dealt out his distressing pangs with torturing severity, but life had still been spared. "The great change," however, was near, and as the morning of the 11th unwillingly lifted up its shaded light, one soul "returned to the God who gave it." This being the first death, it may reasonably be presumed that some account of the last scene will be expected. The deceased, William Curryall,[5] a native of Stromness, was in his fiftieth year, apparently of a healthy constitution, but rather subject to nervous debility. For some considerable time he had been labouring under depression of spirit, and seemed to have given up all hope of relief from the day he was seized with scurvy. As the disease strengthened, his mind got rather weak, and up to the time of his death did not recover its wonted tranquillity. He died without a struggle. As soon as possible the body was carefully wound in a blanket, sewed up, laid out on the carpenter's bench, and then all were requested, who were able to assemble, to join in the last service. This call was willingly obeyed, and each one seemed to feel with deepest sincerity the full force of the solemnities of such a scene. The funeral prayer was read by Mr. Littlejohn, and then the corpse was carried, in mournful procession, to an opening in the ice, through which it was consigned to a watery grave.

On the 14th, the daylight was getting pretty strong, and the sky kept very clear. About noon, several lanes of water were seen, and what was more rare some whales were blowing in them. These were the first that had been seen since the Dee was beset; but the men were unable to make any attempt to kill them, and, besides, it would have been almost, if not altogether, impossible to have done so.

Nothing particular occurred until the 16th. On that day there was joy even in the midst of grief. By the calculations made from the previous observations it was not expected that the sun would be seen before the 25th; but most unexpectedly he appeared on the 16th. This was indeed a pleasing sight, and so delighted was each one when he saw the "orb of day," that half his load of woes seemed to have been removed. The scene afforded matter for very diversified and interesting contemplation. It was pleasing to witness the return of such a friend; but what changeful thoughts must have crowded on the mind while the eye gazed on his span-like altitude, and then followed his early obscurity. Now he seems to rise in all the "glory of his strength"; darts a cheerful ray on the icy regions of space; extends his healing influence as he "exalts in might"; lingers for a moment as if in vision his eye had fallen on the frozen barks, but deeming their ill-fated crews unworthy of his regard, shrinks back from the sight, and shades their sorrows, as if he, too, had conspired against them.

The sun's altitude had hitherto been taken on board the Dee by Captain Gamblin, but now he was unable to do so. For some weeks he had been complaining, and seemed to suffer, too, from depression of spirit. The utmost attention was paid to his wants, and everything done which could in any way relieve them, but he made no progress towards health. Under these trying circumstances, the mate took an observation, and set down the latitude at 69° 1′.

Seeing the crew getting weaker day by day, with no prospect of recovery, the mate thought it would be well to take in two reefs of the topsails. When this duty was ordered, only fifteen men were able to go aloft, and of these the greater part were very weak. The motive which induced the mate to reef the sails, was a conviction of its being exceedingly probable that the effective hands would be so far reduced as to be unable to shorten sail should the vessel get out and a gale come on.

William Besley, from Aberdeen, died on the 19th. The sufferings he underwent were of the most excruciating kind, but he seemed to endure all with patient resignation. The same services were gone through in this case as in the other, and his body was sunk at a considerable distance from the vessel; those who were able attending in mournful procession.

So soon as the announcement of another death reached the ears of the sick, each one gazed in silent thought on his fellow, but none expressed the feelings of his mind. It was often observed, however, that such announcements at first caused much depression of spirit; but, when the deaths had become more frequent, they were less impressive on the survivors.

The ice was now remaining firm, and the winds were light, but the frost was getting more and more distressing, with no appearance of a thaw. During the last week of this month, several whales were seen, some of them very near; unicorns were also observed in the lanes of water; but there was no one now caring about anything but his life, and even life itself seemed a burden. The melancholy tidings of the Captain's hopeless state had a severe effect on the men, and led several of those who were inclined to hope for life now to despair. It was a convincing proof of the mortality of the disease that, even with the privileges the Captain possessed, it could not be stayed; and surely, if in this case death was almost certain, there could be little hope between decks.

Andrew Bennett, from Aberdeen, died on the 27th. He had caught cold by the exposure; and, being of rather a delicate constitution, his lungs were injured. He was also seized with scurvy, which, together with difficulty of breathing, made his case one of peculiar distress. He was quite sensible to the last, and died very quietly.

January ended by a most disagreeable change of weather. The ice again threatened to break up and crush the vessel; but, by this time, many were so tormented with pain that death seemed a consummation rather to be wished than dreaded.

The month of February opened under the most alarming prospects. Three of the crew had already "gone the way of all the earth;" and the fatal disease was making such dreadful inroads on the health of the survivors, that a serious increase of mortality was hourly expected. John Setchell, a native of Hull, but married in Aberdeen, expired on the evening of the 1st. Throughout the voyage, Setchell had been weakly, and his predisposition made his death almost a matter of certainty, now that scurvy was so common. The 2d saw no blank; but, on the 3rd, Captain Gamblin was no more. This was the heaviest stroke, and could not fail to throw a gloom over the already dark and dismal scene. Captain Gamblin's sufferings were most excruciating, and more particularly towards the close of his life. His body was wasted to a shadow, and his well-known manly and powerful voice could only breathe like the whisperings of a child. The nervous debility which had pressed heavily on him during the earlier stages of his sufferings increased to a painful extent a few days before he died; and the thought of his wife and family at home so distracted his mind, that at times he gave vent to the most frantic grief. Compared with this, his own suffering seemed to give him little trouble; and, if there was one motive which, more powerfully than any other, induced an anxious hope of relief, it was the earnest wish that he might be spared to his family. Under this impression, he was incessantly talking about home; and, even when the strength of disease had deprived his mind of its wonted composure, the burden of his soul was his dear wife and little ones. But his hour was come — he died; and, while dying, breathed a last and fervent prayer that a kind Providence would realize, in the experience of those whom he had left behind, the fulfilment of the promise, "I will be a father to the fatherless, and the orphan's stay." There is one circumstance connected with Capt. Gamblin's death which deserves to be particularly recorded. It is this; — it was the sixth hour, and all who were able assembled in the 'tween decks, and within hearing of those who were on the beds of death. Each one was silently and deeply impressed with the realities of the heart-rending scene, and all seemed now more than ever sensible of the necessity of early preparation for eternity. On this side, an empty couch of the dead revived past recollections; and, on the other, the writhings of inconceivably severe sufferings, pierced the soul with deepest sorrow. Thus feeling and thus meditating, the mind was in a peculiar manner disposed to listen to the instructions of the sacred volume. Mr. Littlejohn had just begun the service, a passage from the New Testament was being read; and, while all were attentive and earnest, the death-toll was heard — the spirit of the Captain had that moment fled.

From the peculiar intimacy which had subsisted between Captain Gamblin and Captain Taylor of the Grenville Bay, the mate despatched a messenger announcing the melancholy intelligence. Captain Taylor immediately came over to the Dee; and, on consulting with the mate and the surgeon, it was resolved to preserve the body of Captain Gamblin as long as possible. A coffin was therefore made by the carpenter of the Grenville Bay, the carpenter of the Dee being confined to bed, and the body was then placed on the quarter-deck.

During the two following days, a gale of wind from the east again threatened the vessel, and the danger was the more increased on account of the helpless state of the crew. Every hour some one was taking to bed; and of those who were able to walk about, only some eight or nine could do duty. The drift south, in the course of these two days, was only about eleven miles; but, though the latitude was getting lower, the frost increased, and was much more severe than when they were as far north as 75°. To give some idea of the intensity of the frost at this time, it may be stated that every liquid was frozen; the water in the casks was a piece of solid ice; and, even while the snow was being melted to cook the victuals, the icicles were hanging by the side of the scuttle-lass, or water-cask, distant from the coppers or fire-place about six feet! Every chest between decks was white with frost, both outside and inside; and, as a proof of this, it need only be mentioned that, so soon as a thaw came, some pints of water were in every one of them. With such severe frost there, what must have been the sufferings of those who were not able to come within sight of the fire! How deplorable a sight was it to behold the very blankets in the bed covered with solid ice, especially by the sides of the vessel — the pillows frozen in every part but where the head lay, the very hairs of which were in some cases stiff with cold — each day the frost getting more bitter, and each hour the unfortunate sufferers becoming more and more unable even to break the ice on their wretched coverings. About this time, too, vermin began to make their appearance, and so suddenly did they increase, that in some of the beds they were literally swarming;[6] and these vermin were of a most rapacious kind. In many cases, they found a lodgement underneath the skin, and fed on the flesh like cancerous lechers. But so horrifying would the details in this case be, were they minutely given, that no human being would believe them. Besides scurvy and vermin, the sufferers were almost all seized with most violent diarrhoea; and so dreadfully did it affect them, that relief for a single half hour would have been hailed as an invaluable respite. Under such a complication of disease, it was impossible for the mind to be otherwise than in a state of mental abstraction; and, while this prevailed generally, it was remarked that those in particular who were conscious of having lived a thoughtless life were in the most deplorable state; and, indeed, so awful were some of the cases, that the survivors say their expressions of despairing prospects can never be forgotten.

George Dawson, from Shields, a spectioneer, died on the 5th. In the afternoon, the mate made a calculation of the quantity of provisions, and the probable time of getting clear; and, believing that some additional allowance might be given, he decided on giving it on the bread and barley. This was matter of great thankfulness to those who were able to partake of food, and is believed to have in some measure tended to preserve life which would otherwise have been lost. The spirits had been reduced for some time; but the men generally did not give themselves much trouble about that, no good having been derived from them. But the severest hardship, apart from bodily pain, was the want of fire. The coals had been done since January, and sticks, or staves of casks, were all that could now be had. These were not sufficient to throw out heat so as to dispel the frost between decks; and, what was worse, they were only made use of to cook the victuals. The only other time fire was allowed, was when ice had to be melted for water to drink; and here it may be observed that, when the men were thirsty, they had to wait with patience until the snow was dissolved, then watch the cooling water, and seize the earliest chance of getting a draught before it was again frozen. Had coffee or sugar been plentiful, the water might have been used warm, but such was not the case. A few of the men had yet some coffee and tea, but the greater part had none.

On the three following days there were no deaths, but, on the 10th, Christopher Sutherland expired; and, on the 11th, William Flett. The former belonged to Shields, and the latter to the Orkneys; neither of them were married. On the 12th, John Isbister, Robert Burns, and Alex. Reid, died. The boy Reid expired in the arms of David Gibb; but his death was more the effect of palpitation of the heart than scurvy. This boy died quite sensible and very happy.

The sun's altitude was taken on the 13th, and the latitude marked 67° 32′. The wind was fresh from NE., and the aurora borealis was very clear during the night. A good deal of water was seen pretty near, but the bay ice was very strong. William Muir, one of the men of the Thomas, died on the 15th.

The west land was seen on the 16th, in latitude 63°

33′; and the water appearing to get rather more free, strong hopes of getting clear soon were now entertained. The other vessels were getting now pretty far north. The Advice was as far as twenty miles distant. No communication had been held with the Advice since the end of Dec. At that time, two of the men came over to the Dee, and represented the state of the crew to be so very sickly, that no more on board were able to travel so far had they been inclined to do so. They also stated that the Advice had made a little water, but nothing worth noticing, the crew being able to pump it out. She had got one squeeze; and, so dangerous was her situation, that part of the provisions had to be taken on the ice. Scurvy was the disease from which the men were suffering.

No more deaths happened on board the Dee until the 23d. On that day, John Dunbar died. He belonged to Aberdeen. John Learmond died on the 24th. Samuel Brown, from Shields, the cook, and James Gaudie, both died on the 25th. On the 26th, John M'Leod died; and, on the 27th, John Booth also expired.

March revived peculiar recollections. In previous years, nearly all on board had been busily occupied during this month in making prepartions for the fishing. *Then* they looked anxiously and hopefully forward; but, *now,* how different the prospect! There was, indeed, something peculiarly pleasing in anticipating a return home; but the distressing situation in which the men were now placed deprived the mind of its natural capability of deriving anything like satisfactory pleasure from the prospect; and, besides, how many contingencies did it involve! The season must now be near when the ice would break up; but, then, every day reduced the number of the crew; every hour, those who were as yet comparatively healthy were falling victims to disease; and the probability of no effective assistance being awaiting previous to reaching home — all conspired to darken the hope and dash the expectation.

Alexander Anderson died on the 1st of March. The wind prevailed from E.NE. during the next three days of the month, and the ice became very loose and dangerous. At one time, the probability of a fatal squeeze was so strong, that the provisions were all put in readiness for being taken on the ice; and, while these preparations were making, what must have been the feelings of those who were unable to be removed from their beds?

The ice closed on the 3d; and the foretopsail, which had been set in expectation of speedy relief, was again stowed. On this day, Robert Moir, of the Thomas, died. On the 4th, John Booth, from Aberdeen; and, on the 6th, Andrew Masson, also from Aberdeen, both of the Dee's crew, shared the same fate. On the 7th, James Yorston expired; but no more were cut off during the next four days.

The latitude on the 9th was 62° 49′; wind NW. blowing very strong. By this time, only six hands were able to do duty; and, under the impression that the Grenville Bay's crew were more healthy, the mate of the Dee went over and requested to know if Captain Taylor could render any assistance were the vessel in the open sea. Captain Taylor did not consider that he would be justified in doing so, as twenty of his own men were on the sick list. About this time, full allowance was given in everything except the bread, which was limited to three and a half pounds per week, each man, but it was too late, few were able to eat it. The Norfolk was now about seven miles distant, bearing N. by W. The Advice nearly as far off bearing E.NE.

On the 11th, the Advice was seen with her sails set, and making way through the ice towards the south. The latitude was this day ascertained to be 62° 43′. In the afternoon, Peter Linklater, of Orkney, died. On the 14th, James Muir expired, also a native of Orkney. A great deal of snow fell on the 15th; and at 2, A.M. a heavy swell came on, which broke up the ice. About this time, the vessel was in great danger, some of the ice almost breaking over her. At 4, P.M. the rudder was shipped, and strong expectations entertained of getting early out. William Stirling, of Aberdeen, died during the night, and was the last one whose corpse was sewed up in a blanket and put down through the ice.

The 16th was a dull frosty day, but the ice was breaking up in all directions, and gave other indications of a clear opening being at hand.

Five months and eight days had now elapsed since the

Dee was beset in the ice, during which she drifted 670 miles south; and after such dreadful and unpredecented sufferings, it may safely be presumed that the survivors were glad to get out. But, alas! many of those who had yet hope of seeing home were denied such happiness.

We have now brought the melancholy narrative down to the period the vessel left the ice; but conceiving that a literal copy from the log-book of the subsequent incidents, &c. will be of more importance than a general description, the following extract is subjoined:—

16th.—Got into the water in 62° N. and 59°W. Three boats hanging over the side, not able to take them in. Put the provisions below. Cut the Thomas' boat adrift. David Dinnet died of scurvy. Ship reaching to the south and eastward under three-reefed foresail and close-reefed topsails. At 2, P.M., James Cook, of the wrecked ship, died of scurvy.

17th.—Moderate breezes; ship reaching to the south and eastwards. Took one boat upon deck and secured the others. The men employed clearing the ship. This day contains twelve hours, to commence the sea-log; lat. 61° 56′.

18th.—Moderate breezes, with cloudy weather. This day we take our departure, bearing in lat. 61° 56′; lon. 58°W. The crew in a remarkably weak state.

19th.—Strong breezes, with dull frosty weather. Stowed the trysail, the main staysail, and foretopmast staysail. At midnight, strong gales; wore ship to the south and westward. At 8, A.M. wore ship to the east. At 10, John Moir died of scurvy.

20th.—Strong breezes; ship under two close-reefed topsails and reefed foresail. At 2, P.M. William Paterson died of scurvy. At 8, wore ship to the south and westward.

21st.—Strong gales. The ice in the holds thawing very fast; ship under the close-reefed maintopsail and reefed foresail. At 3, P.M., Magnus Curryall died of scurvy. At 4, wore ship to the eastward. At 4, A.M. John Isbister of the wrecked ship died of scurvy.

22d.—Moderate breezes, with dull cloudy weather. The seven men able to keep watch complain much of pains in their limbs. At 4, A.M. set the foretopsail and trysail and foretopmast staysail; lat. 62° 1′.

23d.—Fine breezes, with dull frosty weather. Several of the crew in a very dangerous state. At 1, P.M. Alexander Noble died of scurvy. At 10, Alexander Garden died of scurvy. Lat. 59° 55′.

24th.—Strong breezes. Ship under two close-reefed topsails and reefed foresail. At 4, A.M. set the jib and trysail. Crew in a very weak state. Lat. 58° 41′.

25th.—Light variable breezes, with showers of snow. At 2, P.M. James Pearson died of scurvy.

26th.—Light variable airs, with calms. All hands employed cleaning the ship below. At this time, have about a quarter of an hour's spell at the pump every four hours. At noon, tacked the ship to the southward; lat. 58° 50′.

27th.—Moderate breezes and cloudy weather. At 4, A.M. strong breezes; stowed the mizen. At 10, strong gales; stowed the trysail, main staysail, and foretopmast staysail.

28th.—Strong gales; ship making a deal of water; half-an-hour's spell at the pump every four hours.

29th.—Strong breezes, with cloudy weather. Ship under her close-reefed maintopsail, reefed foresail, &c. At 8, wore ship to the northward; several of the crew in a very weak state. Lat. 56° 24′.

30th.—Strong breezes, with dull cloudy weather. At 8, wore ship northward. The crew getting weaker every day.

31st.—Strong breezes, with dull weather. At 6, P.M. wore ship to the southward and eastward. Ship makes less water. At 4, A.M. set the jib and mizen. Lat. 56° 18′.

1st April.—Strong gales. At 1, P.M. stowed the jib; at 8, stowed the mizen and main staysail.

2d.—Strong gales, with dull cloudy weather; ship under two close-reefed topsails and reefed foresail.

3d.—Strong breezes, with heavy showers of rain. At 2, P.M. James Tulloch[7] died of scurvy. At 2, A.M. G. Turriff died of scurvy.

4th.—Strong breezes, with thick weather. At 2, P.M. Allan Monro, of the wrecked ship, died of scurvy. At 8, A.M. wore ship to the northward.

5th.—Calm throughout, with a high sea from the E.NE. At this time, only six able to keep watch.

6th.—Light variable airs, with a thick fog. At 7, P.M. William Anderson died of scurvy. At 8, William Turriff died of the same disease. Twelve miles to the SW. for current. Sun observed.

7th.—Moderate breezes, with heavy showers of rain. At 8, Andrew Spence, of the wrecked ship, died of scurvy.

8th.—Moderate breezes, with rain. The six men able to keep watch get so weak as to be unable to break heavy wood. At 8, P.M. wore ship.

9th.—Fine breezes and passing showers of rain. The men get weaker every day.

10th.—Light breezes and hazy. Ship with two reefs out of the foretopsail and one out of the mainsail. Obliged to cut a hole on one of the casks containing the fresh water, to get out what of it was not frozen.

11th.—Moderate breezes, with fog throughout. Several of the crew in a dangerous state.

12th.—Light breezes throughout, with thick weather and rain at times. At 8, P.M. Thomas Stuger died of scurvy. At 2, A.M. John Davidson, of the Thomas, died of the same disease.

13th.—Moderate breezes throughout, with thick weather. The men able to keep watch are six in number, most of whom are weakly; those confined to bed get weaker every day, among whom is Mr. Harris.

14th.—Moderate breezes, and clear throughout. The sick in much the same state as mentioned above. Lat. 59° 23′.

15th.—Strong breezes, with passing squalls and showers of hail and snow. Lat. 58° 54′.

16th.—Moderate breezes, with clear weather. At 1, P.M. David Smith, of the wrecked ship, died of scurvy. Lat. 58° 40′.

17th.—Hard squalls, with cloudy weather; stowed the foretopsail. At 8, worship. Showers of hail and snow towards the end.

18th.—Calm, with clear weather. At 1, P.M. James Isbister died of scurvy. The men able to keep watch begin to get very weak; complain a good deal of their gums and limbs.

19th.—Light breezes, with clear weather. Out the reefs

of the foretopsail, and set the jib and reefed mizen. Lat. 57° 54′.

20th.—Moderate breezes, and clear. At 6, P.M. saw a brig standing to the northward; hoisted a signal of distress, expecting she would bear down upon us, but, to our great mortification, no notice whatever was taken of us. Jib-sheet gave way, and, upon getting it made fast, it split. On clewing up the foretopsail, it split. Ship has made a deal of water during the night; men very much fatigued working at the pumps.

21st.—Light breezes, with clear weather. The crew at the present time remarkably weak. Ship has made less water; if she had continued to make as much as she did yesterday, the men would have soon been completely exhausted. Mended the foretopsail, and set it with three reefs.

22d.—Light breezes throughout, with calm and clear weather. David Irvine, of the Thomas, died of scurvy. Let out the reef of the foretopsail; saw land, bearing E. by N.

23d.—Variable airs, with calm and clear weather. Ship drawing to the southward; land seen to the northward and eastward.

24th.—Light airs and calms, with clear weather. A great many islands to the southward of us; do not know what islands they are. At this time, only three hands able to go aloft.

25th.—Moderate breezes, with clear weather. Saw a fishing boat; hoisted a signal; boat came alongside; learned from the fishermen that we were nigh the Butt of Lewis. At 6, P.M. the barque Washington of Dundee, Mr. Barnett, bound for New York, bore down upon us, and inquired if we wanted any assistance. On informing him that we had only three hands able to go aloft, he sent four men to our assistance; came on board himself, bringing with him wine, porter, &c.

26th.—Light breezes, with clear weather. The barque took our ship in tow, and towed her to the Sound; committed her to the care of a pilot. In the evening, the surgeon appointed by Government came on board to visit the sick. Anchored in Strómness harbour about 11, P.M.

27th.—Light breezes, with clear weather. Ship at

anchor in Stromness harbour. Early in the morning, the surgeon came on board, and got the sick conveyed on shore to the hospital. Several men sent on board to do sundry jobs.

The most remarkable incidents peculiar to the voyage from the ice to Stromness were the refusal of the first vessel seen to render assistance, and of the crew in the boat to come on board. It were premature to condemn the conduct of those on board the ship, as it may have been that they did not perceive the signals, but the men in the boat are not so excusable. The paragraph in the foregoing log, which refers to the boat coming alongside is not so explicit as the following, which was noted at the time by one of the seamen:— "This morning, at six o'clock, we saw a boat containing eight men; we hove to, close to the windward of the boat; we waved upon them to come alongside; they came and asked us in English what we wanted. We told them that we had been all winter in Davis' Straits; and that, if they would come on board, and assist us to Stromness, they would be well paid for it. They would not come on board. We hove two pieces of meat into their boat; the boat being half loaded with fish, we hove down a rope to them, and asked of them to give us a fish, but they refused, and pushed off." Now, we cannot conjecture what could have been the motives which induced these fishermen to refuse assistance; and not only so, but to forego the prospect of a rich reward. Were it not that we know British seaman better than to believe that they could be callous or indifferent in such a case as this, we should condemn these men at once as a set of merciless scoundrels. But, when we think on the sickly appearance of the men on board the Dee, the coffin on the quarter-deck, and other forbidding circumstances, we are inclined to believe that the fishermen suspected a case of plague, and thus refused to put themselves in a situation where they would most likely fall victims to its dire influence. But the taking of the pieces of beef and refusing to give a fish in return, are drawbacks on this opinion; and, if the men can reconcile such conduct with the claims of suffering humanity, they are more objects of pity than of hatred. By order of the owners, Captain Goldie, from Aberdeen, went

Sailor's model of the *Swan* of Hul

A whale's jawbones arch a Stromness gateway in the ea 20th century.

An Orkney whaler's pay warrant.

STROMNESS, 13 march 1816

£0. 15. 0 *per Month.*

Two Months after date, please to pay to Nicol Seth *or Bearer hereof* fifteen Shillings *in part of his Wages as* Seaman *on board the Ship* Thomas *now bound upon a Whale Fishing Voyage to Greenland or Davis' Streights, and back, and continue the payment of the said sum monthly until the completion of the aforesaid Voyage, unless sufficient reason to the contrary shall appear.*

Geo. Geddes

To Mr. George Geddes,
STROMNESS.

to Stromness, and, having engaged effective hands, brought the Dee to Aberdeen.

There is another circumstance, in connexion with the voyage from the ice, which is worthy of particular notice. The men who were spared were inclined to hope that some Government vessels would be cruising in search of the missing whalers; but, having fallen in with none, though sailing in such wide latitudes as the vessel did during the first few weeks after she got clear, the men got low in spirit, which, perhaps, tended to hasten the melancholy fate of those who so soon thereafter died.

The Dee was not much damaged as might have been expected, and, indeed, looked much cleaner than many vessels after an ordinary voyage. She got some cleaning at Stromness; but, even when she arrived there, she was not damaged to any extent worth mentioning. When the foregoing details are read, this circumstance will appear rather singular; nevertheless, such is the fact, and it can only be accounted for in the great strength of the vessel.

A good deal of provisions were brought home; and some persons have been uncharitable enough to throw blame on certain parties, whose conduct, they think, in reducing the allowance to one-half so early in the season, is questionable. This is a rash condemnation; it need only be known that the Dee was very early beset, and in a latitude where there was every probability of a long wintering, to do away with an suspicion as to the propriety of reducing the allowance. There is a question, however, which cannot be so easily disposed of. Why did the owners, in the face of last year's experience, not put a twelvemonth's provisions on board? Of course, nobody could have forced such additional provisions, and therefore the owners are not legally blameable; whether they are *morally* so is another question. But the fact is, Government should render such additional provisions absolutely binding on those who are connected with the Davis-Straits' fishery, now that the hazardous nature of the trade is so very obvious.

The case of the Dee, too, shows the necessity of a very full supply of medicines suitable for the scurvy. Mr. Littlejohn was perhaps well prepared; but there is reason to think that had his chest contained a larger quantity and a greater variety, the disease might have been mitigated.

The seamen who take out coffee and sugar at their own expense, will see, from this trying case, that they ought to be careful in laying in a large stock. Had those on board the Dee been amply provided with these articles, they would have felt the benefits of them when they had nothing to drink but water, half frozen, though newly melted from the snow.

The Dee arrived in Aberdeen on the 5th May, which is thus noticed by the *Aberdeen Herald:—*

"The Dee arrived in the bay on the morning of the 5th, and at noon entered the harbour. The quay was crowded with anxious spectators; and, as the vessel neared the berth, the scene was truly heart-rending. The mourning relatives of the deceased seamen, though previously apprised of the unfortunate fate of those who were near and dear to them, seemed unwilling to give credence to any testimony apart from a positive confirmation by those who had been eye-witnesses to their decease; or, believing the fact, seemed anxious to seize, with eager avidity, the earliest opportunity of taking a parting glance at the empty hammocks of the dead. Their weeping widows rushed on board with their helpless orphans in their arms, while parents and friends followed in equal grief. Of those who were privileged to meet their surviving relatives we need say nothing — their joy was great, but the detention of a few who were left at Stromness, led the expectant friends to give vent to the most frantic grief, and almost again to despair."

The total number of deaths on board the Dee was forty-six; of these, nine belonged to the Thomas of Dundee. Since the arrival of the Dee at Stromness, one of the invalids has died, thus making the loss no fewer than forty-seven souls. The numbers of widows and orphans left is very great, and their helpless condition renders them objects of great commiseration. It is to be hoped that a benevolent public will extend their liberality very freely in their behalf.

ACCOUNT II

The Grenville Bay

In 1985 the typescript of this account was received by Stromness Museum from Miss Amelia Peterson of Polegate in Sussex. It was accompanied by a letter from which this is an extract:

> The following log was copied from the original writing in an old exercise book which belonged to my cousin Jessie S. Ritch with whom I lived. It is believed that the writer was her forebear.
>
> My cousin died in Dec. 1951. After making a copy of the log to which I felt entitled as her private books and papers came to me, I returned the log to her Aunt Margaret Ritch [in Deerness] as I felt that it belonged to the Ritch side of her family in Orkney. We were cousins brought up as sisters, on our Grandmother's side.
>
> Margaret Ritch, her father's sister, died in July 1954. After I returned the book to her, she wrote me a letter saying: 'I remember one Deerness Man living that came back from the ice. He was beadle in the church at the old school. I think it was the Grenville Bay they called the ship. He used to sleep in church and the precentor always kicked his desk to waken him.'

EDITOR'S NOTE

The original log cannot now be traced. It was, however, published in *The Orcadian,* 24/12/1931, and then reprinted as a pamphlet. There are minor divergences between that and Miss Peterson's typescript. Mostly I have followed Miss Peterson's version, but in some places the *Orcadian* wording is clearly preferable. The original was clearly lacking in punctuation. Spelling has been tidied in both accounts, but occasional errors survive. The authors would have spent many years less in school than today's pupils. The entry of 23rd December has survived intact in Miss Peterson's manuscript. The simple sincerity of alarm and thankfulness would be damaged rather than aided by any adjustment to the punctuation.

REMARKS ON BOARD THE SHIP GRENVILLE BAY OF NEWCASTLE ON TYNE

KEPT BY

ROBERT WILSON AND THOMAS TWATT

SEAMEN ON BOARD SAID SHIP

This journal is kept from the 1st October 1836 to the 27th April of her drift through the ice and her passage home to Orkney. This log is kept as a harbour log from 12 o'clock at midnight until 12 o'clock next night.

SKETCH OF THE VOYAGE TO DAVIS STRAITS OF THE SHIP GRENVILLE BAY OF NEWCASTLE ON TYNE

We joined the aforesaid ship at Stromness April 15th 1836 and lay there wind-bound until the 29th. We then put to sea with the wind S.S.W. but had not sailed far till the wind came nigh ahead of us and blew very heavy, so that all hands were called to shorten sail. We close reefed our top sails but the wind still increasing we were obliged to heave to under a close reefed main top sail. We lay in that position for twelve hours before the gale began to abate. The sea was very heavy but we received no damage from the violence thereof, only some of our crew were very sea-sick. The wind became more favourable and we made all sail to proceed on our journey as quick as possible, but the weather was very changeable, and often blew very heavy from the Eastward that we could not run and consequently were very much tossed about until the 15th of May when we reckoned ourselves to be about seven hundred miles from the British land.

This day the 15th of May a fatal accident occurred. We had a very strong breeze from the N.W. and on the 15th the wind moderate and in the afternoon the wind shifted round to the Southward. We then began to make sail, but it fell almost calm and a heavy sea from the N.W. At seven in the morning we began to make sail. We set our fore topmast studding sail and lower studding sail and three of our men went up to set a main topgallant studding sail and one of them went to reive the halyards in the block at the mast-head. He lost his hold and fell on the deck without touching anything. This was a dreadful sight to all beholders. The doctor was called immediately and every possible means was used to restore animation but all in vain. Life was extinct. This took place on Sabbath the 15th May at nine minutes past 8 o'clock in the evening and on the 16th at 6 a.m. his remains were committed to the deep.

We shortly after got a fair wind which brought us near to our destination, and the weather looking favourable we put out all our boats, and sent up our royal masts and yards, and rigged our ship in the country fashion, but the day after we had a heavy gale of wind which almost laid us on our beam ends and was nigh to have lost two of our boats in consequence of which we were obliged to call all hands to take in our boats and to send down our royal yards and masts and also our top gallant yards and masts. We then proceeded up the country until we came to the S.E. Bay where we fell in with a heavy run of fish but could not succeed in catching any of them. We then went back to the S.W. to Lat. 62°30′ but saw very few fish. It was now about the middle of July and we were still clean. We then went up the country to try for a passage to the West land and when we came to the Thumb[1] we met a number of vessels there and the ice was close as far as we could see and consequently there was no prospect of our

GREENLAND WHALE

getting a passage in consequence of which all the captains agreed to go South. After this management the ship *Margaret* of London came up and told us that they had seen the North water and that we might get a passage quite safe.[2]

We then put about and came to the place and got into the water a few hours after and fell in with a number of large fish and caught one of them. We then made sail for the Westland ice and when we came there it had a very formidable appearance and no prospect of a passage to be had this year it now being the latter part of August. We immediately put back to try to get to the Southward being the same way we came. We fell in with a few large fish on our way and got one of them. She stove one of our boats and upset another[3] but thank God there were no lives lost. We made sail again and came to the place we before crossed but finding it entirely blocked up we now began to dread our sad fate. There was no vessel in company with us but the *Norfolk* and consequently were afraid to proceed so far being alone. We then went back to the Westland ice to see if we could fall in with more vessels and when we came there we met in with *Dee, Thomas* and *Advice* and made known to them our sad case which caused them to put their men on allowance. We had been put on allowance before. We then made an effort to get Westward but of no avail. The N.W. wind appearing to prevail we again made sail to the Eastward. The *Norfolk, Dee* and *Advice* and *Thomas* in company but the *Thomas* and *Advice* being fast sailing vessels they arrived there long before us. When we came there and could not see them we suspected they had got through and made all possible endeavour to follow them but of no avail.

No tongue could express our feelings now all hope of our reaching home were gone and nothing to expect but a dark dreary winter excluded from all the fond endearments of a cheerful home. It was now agreed to go South and to keep burning blue lights on the night time to keep the ships from separating and when we came to the Westward as far as we could get we met in with the *Thomas* and *Advice* there before us and were very glad to find them. There was now a meeting held and it was now agreed that we should proceed to the Southward as far as possible and make the ships fast, and saw docks in the ice for the winter. But now some of the crews were rather rebellious and threatened to take charge of the vessels from the masters and some of our crew seemed to loose all reason and where ever there were ten men together there were ten different opinions everyone thinking of his own the best. Now our melancholy situation was full in view. To our own imagination nothing cheerful could present itself. Having nothing before us but darkness, severe cold, hunger and death. May our confidence be placed in Him who is able to deliver us out of the greatest dangers. Although now in the remotest part of the earth we are not beyond His powerful preservation whose eye seeth in every secret place and will not forsake them that earnestly seek Him. Upon the 1st day of October we all made sail standing to the Eastward to a crack we saw in the floe where we shall begin our journal.

A SERIES OF REMARKS

Sat. Oct. 1st. Strong gales and cloudy. At 8 a.m.[4] Made sail standing to the Eastward. The *Norfolk, Dee, Thomas* and *Advice* in company. At 1 p.m. all the captains went on board of the *Dee* and agreed to stop in the bight for which we were sailing.

2nd. Strong gales and cloudy. The ships all in company. At 10 a.m. the *Thomas* and *Advice* made all sail determined to proceed Northward till they got round the end of the ice. The *Dee* and *Norfolk* followed them. Our captain then called all hands to know whether they would be agreeable to stop in the bight alone or go with the other vessels. They answered with the other vessels.

3rd. Fresh breezes and cloudy, turning North the same vessels in company. Our captain called all hands this day desiring them to go upon less allowance of meat but they would not consent to it as yet.

4th. Fresh gales and cloudy weather, turning North, the *Norfolk* and *Dee* in company. Ice in sight to the Eastward. Wind E. by N.

5th. Light winds and a heavy fall of snow. This morning we espied the *Thomas* and *Advice* bearing North from us. At noon calm, all the ships fast in the pack.

6th. Calm and cloudy weather all the ships fast in the ice.

7th. Calm and cloudy. At 10 a.m. moved a little to the Southward. The *Norfolk* and *Dee* close in company. The *Thomas* and *Advice* 3 miles North of us.

8th. Light winds and clear. At 2 a.m. the moon and stars very bright. The bay ice making very fast. At 7 a.m. made sail going a little South a boat employed ahead breaking the bay ice.[5] At 4 p.m. hoisted up the boat. Lat. 73° 15′ N.

9th. Light winds and variable. At 8 a.m. lowered down the top sails. At 11 a.m. the *Dee* set her top sails making towards us. At noon all the ships fast. Frost very severe a little snow falling. Lat. 73° 15′.

10th. Light winds and variable. Cloudy weather. at 7 a.m. the sun rose above the horizon. At 8 a.m. the *Thomas* and *Advice* made sail making a little towards us. Distance from us about ½ mile bearing S.E. This day at 9 a.m. we were put on short allowance of bread and beef. Our allowance now per week being 4lb. of bread per man and a proportional allowance of other things.

11th. Light winds and cloudy weather frost very severe. All the ships remain in the same position as before. At 8 a.m. sent a watch of men over the ice with two warps to meet the *Dee's* men with other two making them fast and heaving them tight to prevent them from separating. Wind East.

12th. Light winds and clear weather. The ships still fast as before. Strong frost. The ice six inches thick. At noon laid our decks over with coal tar, sand and sawdust over our bed cabins to prevent the frost from striking through. At 7 p.m. stowed all our sails. A little snow falling. Wind E.N.E. Lat. 73° 7′.

13th. Light winds and cloudy weather. At 6 a.m. discovered a crack in the floe between us and the *Dee* about 50 yards broad. Captain Gamblin took a watch of his men and launched a boat to the crack. The captain and three of his men came across to us to talk of our melancholy state. Several cracks appeared. Frost moderate. Wind E.N.E.

14th. Light winds and dull weather. Frost very severe. This day thermometer 20° below freezing point. At noon unshipped our rudder. The wind variable. Ships all fast as before.

15th. Moderate winds and dull weather.

16th. Sunday. Stormy dull weather. At 10 a.m. retired to Divine service in our cabin while the tears of sorrow bedimmed our eyes when we thought on our former enjoyments compared with what we now saw before us. Lat. 72° 54′. We are drifting South, but there are dangerous nips among the ice in different places which are alarming to us at present. Last night the *Norfolk* was visited by a bear.

17th. Moderate winds and dull weather attended with snow. At 7 a.m. observed the *Dee* to heave athwart with a heavy press of ice turning up her side. The crew was employed in getting their provisions on the ice. She was about 60 yds. distant from us. At 8 a.m. we called all hands to take our provisions on deck and get 6 tackles all ready up to be in readiness for lowering them on the ice if necessity so required. We then put in a third tier of beams in our ship and pumped off our second tier of water on purpose to lighten and strengthen her. The other ships employed in taking ice on board for water. Wind N.E. by N.

18th. Strong gales and dull frosty weather attended with snow. Several cracks around us. A lane has broke out between us and the *Norfolk*. The nip is eased from the *Dee*. Wind N.N.E.

19th. Strong winds. At 8 a.m. called all hands to go with two boats to a berg about three miles off to fetch ice to dissolve into water. Our carpenter and his mate employed in making a sledge for dragging ice. At 2 p.m. took our coppers aft in our half deck. We have our ship all in readiness for the winter. Our crew very low in spirits. At 11 p.m. the *Dee* and *Norfolk* had a heavy press of ice forcing all hands to turn out of their beds to go on the ice. Frost very severe.

20th. Strong gales attended with snow. At 2 a.m. the flow split ahead of the *Norfolk*. She set her topsail and run down the crack at S.W. All the vessels have a very heavy press on them this forenoon. The *Dee's* men have all their clothes on the ice. Wind S.S.E.

21st. Strong gales and clear weather alternately attended with snow. We are employed in lining our bed cabins with old canvas, and screening in our mess berths to prevent the wind as much as possible from entering therein. Frost moderate this day. Lat. 72° 35′. Wind S.S.E.

22nd. Strong gales and clear weather. At 6 a.m. the *Dee* had a heavy press again which caused the crew to retire to the ice. All the ships in great danger at present. At 8 a.m. the *Dee's* crew began to saw alongside of their ship. At 4 p.m the ice eased. Frost very severe. Lat. 72° 26′ wind S.S.E.

23rd. Sunday. Strong gales attended with snow. The *Dee's* boats and chests remained on the ice as yet. At 10 a.m. retired to Divine worship. In the afternoon the *Norfolk* had a press on her which caused them to call all hands to saw. The ice has been turning astern of us all this forenoon. At 6 p.m. it came close to our stern and continued coming till midnight turning up our side all the while which caused us great trouble of mind. We called all hands to try what could be done but found it impossible to do anything. The *Dee* and *Norfolk* had presses on them at the same time. Men travelling from one ship to another to know if we were still safe. We had a very dreary night of it neither fire nor meat, and sleep we could not allow ourselves. Wind N.E.

24th. First part of this day now moderate, but towards night a heavy gale from the westward. The press has eased from the ships. We are employed in taking things on board again. The *Norfolk* and *Dee* are in a crack. Our ship is fast in the same ice, but no press at present. We sent our mizen top gallant yard and mast down and took what bread was in our lockers and put it in a cask to be more ready if required.

25th. Strong wind and dull frosty weather with snow. At 8 a.m. the *Dee* called all hands to saw a Dock. We sent a watch of our men to assist them and hauled her in at 4 p.m. Wind S.S.W.

26th. Calm and cloudy. At 8 a.m. all hands were called to fetch ice from a berg 4 miles from us. Returned at 4 p.m. Frost moderate.

27th. Light winds and dull weather with snow. At 8 a.m. call all hands to fetch ice from the same berg with our boat. At 2 p.m sent down our fore and main top gallant yards and masts. While we are thus employed in preparing our ship for the winter no tongue can tell our feelings all expectations of home being now over. Wind S.S.W.

28th. Light winds. At 8 a.m. called all hands to bring ice from the same berg. We are employed every moderate day in the same way before we lose sight of the sun which is going fast from us. This morning several lanes of water to be seen.

29th. Moderate winds attended with snow. At 8 a.m. called all hands to fetch ice as before, but by appearance of the ice and a heavy fall of snow we were compelled to drop it. At 10 a.m. the *Norfolk* called all hands to saw and we sent a watch of men to lend them assistance. At 11 a.m. made a signal to the *Dee* for assistance. We were all employed in sawing and sinking the ice until 3 p.m. We observed the *Thomas* and *Advice* people to be sawing likewise for the safety of their vessels. Wind W.S.W.

30th. Sunday. Strong gales attended with snow. At 8 a.m. the *Norfolk* called all hands to saw again. At 10 a.m. we retired to Divine service. We have seen several unicorns in cracks this day. in the afternoon we were visited by two men from the *Advice* who informed us that the *Thomas* had shot 3 bears. We have very little strange news to expect at present but hope that the ships will all keep safe. Wind E.N.E.

31st. Light winds and clear weather. Captain Taylor had a lunar observation last night and to our sad disappointment we have found that our ship had driven 30 miles North since last observation by reason of the wind being from the Southward. Lat 72° 42′. Winds variable.

November 1st. Light winds and dull weather. Two of

our men travelled to the *Thomas* this afternoon about two miles from us. This is all we can do at present for any amusement and this we cannot have long because our daylight is but small at present. Wind E.

2nd. Dull weather. At 8 a.m. called all hands to fetch ice. With hunger and cold scarce able to travel so far and less able to work hard. Wind variable.

3rd. Light winds and cold weather. At 8 a.m. called all hands to fetch ice from the same berg. This is hard work for us at present by reason of the cold weather and our low diet, but we do not expect to fetch much more as snow will answer the same purpose. Frost very severe. Wind W.

4th. Stormy and thick snow drift. All this day the ice has been turning up in a dreadful manner ahead of us. At 1 p.m. the *Norfolk* called all hands to saw and a watch of our men went to assist them and was with them until we observed the floe come close to our own ship. We then went to our own vessel and found the ice had come to her in a very rapid manner turning up as high as our haws holes which was no pleasant sight to us. So much so that it came to our side turning up above our main channels making the vessel to tremble which caused all hands to have clothes in readiness. God alone knows what would have become of us had our ship gone at this time, the wind being so high and the snow drift so thick that we could not have survived on the ice for any length of time. The floe went over a great deal of our fresh water ice which was alongside. The snow drift is so thick that we can see but a short distance. It is impossible to express the hardships that we endure at present. Our only comfort in Him who is able to save alive in the greatest dangers. At 10 p.m. the ice was still and we retired to rest.

5th. Moderate winds. Called all hands. Sent one watch to saw with the *Dee* and another with the *Norfolk*. The *Thomas* and the *Advice* in the same position as before. We have seen the sun this day. We had not seen him for 11 days back. He sank below the horizon at 10 minutes past 1 o'clock. We do not expect to see him many times more until the turn of the season if so that it should please God that we survive so long. Lat. 72° 35′.

6th. Clear weather. At 8 a.m. retired to Divine service in our cabin. Our ship remains in the same position with the ice fast all around her. We do not expect that there is any water down to her keel. Our Captain visited our half deck this evening and read several chapters in the Bible. Wind N.E.

7th. Light winds and clear weather. It appears we are driving South. By our observation last night makes our Lat. 72° 9′. Frost very severe.

8th. Light winds. Captain Taylor had an observation this morning from a star which makes our Lat. 72° 6′. We saw the upper limb of the sun this day at noon. Saw three bears but did not catch any of them. At 5 p.m. saw a fox close to our ship.

9th. Strong gales with snow from S.W. by W. It has been very bad winds for us ever since we came with the ice. It is now six weeks since we were beset in the ice and have only been driven one degree during that time towards the south. Yet we are 250 miles North from where the ships were beset last year.But we hope the winds will be more Northward in the spring.

10th. Moderate winds and a little snow. At 9 a.m. the *Dee* called all hands to saw and a watch of our men were sent to assist them. We were thus employed until 3 p.m. We then hauled the *Dee* in her dock and retired to our own ship much fatigued. Frost moderate. Wind N.E.

11th. Strong gales and clear weather. The *Norfolk's* dock broke up which made them call all hands to saw. At 10 a.m. a watch of our men was sent to assist them in sawing and sinking the ice. The frost very severe. A great many got frost bit and were obliged to go on board. This day at noon the sun appeared in the horizon. Wind E.N.E.

12th. First part of this day light wind and clear weather. This being old Hallowday brings to our view the appearance of a dreary winter. We will not see the sun again until he turns to come North, and God only knows if we shall be spared so long or not. All the masters visited the *Norfolk* today. The latter part of this day strong gales attended with snow from S.S.E.

13th. Strong gales and clear weather. This morning there was a fox close along side, but we did not catch him there being no gun load. At 10 a.m. retired to Divine worship. This day we were obliged to have lamps burning in our cabin at noon. This evening our captain read several chapters in the Bible to us in the half deck. This evening blowing hard. Wind N.E.

14th. Strong gales and clear weather. This morning our captain had an observation from the morning star which shows that our ship has driven 18 miles south since yesterday morning. Wind N.E.

15th. Moderate wind and clear weather. Captain Taylor has commenced a school for reading, writing and arithmetic and navigation for all those who desire to acquire a knowledge of these branches of education. The *Dee* shot a fox last night and another this morning. Frost very severe this day. Lat. 72° 13′.

16th. Light winds and clear weather. This day we cannot see to read on a book with our hatches off without the help of lamps. The latter part of this day strong winds from the S.S.W.

17th. Strong gales attended with snow from S.S.W. We suspect that our ship is driving North by reason of the wind being from the Southward.

18th. Light wind with snow. The weather is dark and gloomy at present, but we will soon have the moon which will be a great comfort to us. This is the tenth day of her age. We have driven seven miles north which we ascertain by our Latitude being 72° 20′. Wind N.W.

19th. Strong winds and dull weather. We have been employed in screening the main deck over to keep the wind from us as much as possible for a man cannot walk the deck when the wind is strong without the frost catching hold of his face. Frost severe.

20th. Strong winds. We have had no observation these two days but we hope we are driving south as we have strong winds from the Northward. Our master visits our half deck three times every week and reads different sermons and some chapters in the Bible, and comes every night to instruct his scholars, but it is generally to be lamented that there are a great many quite careless about learning. There is scarcely a boat steerer in our ship that knows the alphabet. We now have the moonlight night and

Devil's Thumb', a sketch by Elisha Kent Kane.

Greenland Fiord', a painting by James Hamilton.

day which is a great comfort. Our daylight is like a dark twilight now and will always be getting less. Our earthly comforts are very small at present.

21st. We are employed this day in cleaning the snow from off our deck underneath the covering. Our decks are covered with sails from the foremast to the mainmast, but the snow is getting so much on the top that we are afriad that it will break down. Wind N. by W.

22nd. Clear weather. Last night our captain had an observation which makes our Lat. 72° 7′. Wind S.W.

23rd. Light winds. Ships all fast. We have 20 hours of night but the moon and stars are very bright sometimes. Wind S.S.W.

24th. Moderate winds and dull weather. Saw two foxes this afternoon, foxes and bears are all the strangers we see. Several lanes and cracks in sight this day. Wind S.S.W.

25th. Light winds from the south and clear weather. A large lane of water in sight this day extending from S.E. to N.W. about a mile long. Several cracks to be seen this day. Wind E.N.E.

26th. Light winds from the south and clear weather. At 4 o'clock this morning our captain had an observation which makes our Lat. 72° 1′ N.

27th. Calm and dull weather. At 6 a.m. the mate shot a fox which was a fine fresh soup for the cabin. The wind has shifted to the N.N.E. blowing strong.

28th. Strong winds from the East. At 10 a.m. the captain had an observation from the morning star which makes our Lat. 71° 57′. The ice is squeezing us this day in different places around us making a terrible noise which is very alarming to us, but we know there is a God able to preserve us although placed in an awful situation. Frost very severe.

29th. Severe cold and clear weather. The ice is still squeezing up. Lat. 71° 54′. Wind South.

30th. Light winds and clear weather. It appears that our ship has driven 2 miles North by reason of the wind from the South yesterday. We have been a long time in the ice and driven but very little South. Lat. 71° 56′. Wind N.N.W.

ull whalers *Swan* and *Isabella*. *Swan* can be identified by her figurehead.

December 1st. Strong gales and thick snow. Saw a fox this morning close to our ship. This day we launched our boat to the ship, the boat which was carried off by the ice from along side 3 weeks[6] ago. Wind South.

2nd. Moderate and dull weather. Saw a fox between us and the *Norfolk*. In the afternoon the *Dee* shot two foxes. This evening strong wind from E.N.E.

3rd. Strong gales and thick snow drift. In the afternoon our captain called all hands concerning some disputes that took place last night in our half deck about the bad usage of some of our Orkney men. Our captain made a fine speech to us and told us that all hands should have equal usage for all time coming.[7] Wind E.N.E.

4th. Strong gales and cloudy weather. Snow drift still very thick. At 10 a.m. retired to Divine service. We have only three hours twilight at present. Frost very severe. Wind E. by N.

5th. Strong wind and dull weather. This has been the darkest day which we have yet. Heavy snow drift strong frost. Wind N.E.

6th. Strong wind and dull weather. We have no observation since Thursday last but hope that we are driving south by seeing strange bergs. In the afternoon there were three foxes alongside. The *Dee* has shot six foxes.

7th. Light winds. this day the floe split round us and we are in the middle of a place about ½ mile broad, but it will soon be frozen over again for the frost is very strong. This day we can scarcely look to windward by reason of the keenness of the frost. Our situation at present is very trying but thank God there is yet hope. Wind East.

8th. Fresh breezes and clear weather. We have had no observation these eight days but we hope that we are still driving south for there is a great motion in the ice and some new bergs in sight. Wind E.N.E.

9th. Fresh breezes from the N.E. and clear weather. A very heavy frost rime falling. All the ships fast as before.

10th. Light winds and clear weather. This morning Captain Taylor had an observation from a star which shows us that we have driven 5 miles south every day since our last observation which was ten days ago. Our Lat.

today 71° 2′. Our Captain called all our officers into the cabin to know if they would consent to have one piece of beef taken off the ships company daily which would leave only one piece to each watch in the day but they would not consent thereto and told him to try another week on the same as before; that the present allowance was little enough to support nature. Our boatswain told him that if we were still in confinement till our beef was done that there would be few to eat it. Frost very severe today.

11th. Fresh breezes and dull weather. At 10 a.m. went to Divine service. Wind E.N.E.

12th. Light winds and dull. At 1 p.m we observed the *Thomas* to be listed over by a very heavy press of ice. At 2 p.m sent 3 men to the *Advice* to see if the *Thomas* was safe. They returned and informed us that she was safe as yet but nevertheless exposed to great danger by reason of the ice having lifted her up 3 feet. This is trying to all beholders but more so to the poor fellows that are like to lose their only place of safety for the preservation of their lives. Poor shelter to be had upon the frozen elements. Wind N.E. by N.

13th. Fresh breezes. This morning our captain got an observation from a star which makes our Lat. 70° 24′. At 11 a.m. we observed the *Advice* hoist her ensign to let us know that the *Thomas* was lost which was melancholy news to us as we were all exposed to similar danger and knowing how soon it might be our own fate. At 1 p.m. our Captain called all hands to see if any would volunteer to go to the *Advice* this night and stop there till next morning to be in readiness to assist the wrecked men, as the *Advice* was 2½ miles nearer to the wreck than we were. 21 of us volunteered to go and had much difficulty in getting there as the ice was broken in many places.

14th. Light winds and clear weather. 21 of our ships company and 21 of the *Dee's* were on board of the *Advice* last night and a poor night we had. We left the *Advice* at 8 a.m. to go to the wreck and when we came to the tent where the wrecked men were and what provisions they had saved a mournful sight was presented to our view. There was one man drowned in the night and another lay dead in the tent with part of his hands burnt off in the fire[8] and a great many more in a terrible condition with frost in their feet and hands.

This morning all the ships companies assembled to the wreck and were employed in dragging the provisions in the boats but we could only get it half way to the *Advice*. This we found hard work. The ice is very open and several of us fell in. The wreck is 2½ miles from the *Advice* and three miles from us. We returned to our vessel at 6 p.m. being much fatigued and some frost bit.

15th. Fresh breezes and dull weather. At 6 a.m. all the ships companies went again to try if we could get the provisions to the *Advice*. We accomplished our design but with great trouble. The wrecked men are divided among all the ships. 12 men to each ship their clothes are on the ice yet. We returned to our ship at 5 p.m. Frost very severe. Wind N.E.

16th. Light winds and variable. At 8 a.m. called all hands to go and fetch the wrecked men's provisions to our ship. We arrived at 4 p.m. Some of the wrecked men reached our ship this night. They have not been in bed since their ship was lost.

17th. Light winds and clear weather. At 8 a.m. called all hands to assist the wrecked men to get their chests and clothes to our ship. The *Dee* and *Norfolk* people similarly employed. This has been hard work for us by reason of the severe cold and having to travel so far on the ice, the wreck being seven miles from us. Wind South.

18th. Light wind and clear weather. Aurora Borealis shining bright. At 10 a.m. retired to Divine service. Several of the *Thomas'* men frost bit both in faces and hands and feet and several other diseases attending them by reason of their being so long on the ice after their ship was lost. Wind North.

19th. Light winds and clear weather. The frost is getting stronger every day and we suspect that it will be worse when the days lengthen which will not be long now and we hope please God that we will yet see our native land once more for with Him all things are possible. This is a very dull time with us and our crew is losing heart very much. Frost strong. Wind E.N.E.

20th. Light winds and dull weather. There is a number

of the wrecked men out of health and a number of the *Dee* men in the same state with various diseases. There is a large berg come in sight to the S.W. of us. The ice is squeezing up this day in all directions. At 6 p.m. the ice split along side of us leaving our side bare from the gangway right round our stern. Captain Taylor immediately called all hands to get ready for such a case of emergency. We took the covering off our main deck lowered down our boats and launched them on the ice and got all ready for lowering our provisions on the ice if necessity so required. The crack is still opening and we are afraid of its coming rapidly together for our ship is fully exposed. Wind N.E.

21st. Light wind and variable. The crack is still open only it is covered with bay ice. Our ship is still in danger, but we still trust in Providence.

22nd. A heavy storm from the S.W. and very thick snow drift. We are sore afraid of the ice coming together as the wind is so strong. We lowered down our stern boat and launched her on the ice. We then observed the ice coming together which made us all turn out of our beds. The ice kept closing all day. At 4 p.m. it came very rapid when all were again called to save ourselves but before they could get on deck it came so rapid that it caused some of our men to throw their beds on the ice, but the ice stopped again and all hands went below but the captain. The snow drift was so thick that we had not been below half an hour until we were again called up to save our provisions for the ice was coming towards us in a dreadful manner. We hoisted a light to the *Dee* and the *Norfolk* for assistance. We then began to cut roads in the snow to roll away our provisions from the ship. While we were thus employed our Captain fearing that our ship was gone called us on board to try the pumps and to our satisfaction we found that she had made no water. The ice was thrown up at our stern as high as our cabin windows. At 11 a.m. the wind moderate and the ice was still. Thanks be to God for His almighty power for preserving us from the imminent danger to which we were exposed.

23rd. At 1 a.m. we retired to rest in our beds without we have had no rest these three days. We cannot express the trouble of body and anxiety of mind that we have experienced these three days. We hope that those of our friends who are enjoying the comforts and pleasures of a home in our native land will implore a throne of grace in our behalf for we are in a miserable condition no meat to satisfy our craving appetites and no fire to warm our freezing limbs nor could we have any rest for our wearied bodies. May that God who heard the prayers of Jonah out of the belly of the whale hear us in this remote corner of the world. Our chests and clothes still remain on the ice.

This morning we heard a whale blowing close to us in a hole on the ice. In the afternoon our provisions clothes and boats are covered over with snow and we are consequently employed in digging them out and in digging a road to the vessel. Wind W.N.W.

24th. Fresh breezes with snow. The ice is still as yet. We have all been employed in bringing our provisions on board again. Some of our men went to the captain and told him that their bread was all done by reason of being so disturbed last week. The captain agreed to give us three biscuits each man, it being Christmas Eve.

25th. Light winds and variable. This is Christmas Day and but a dreary day to us, Yet thank God we are all in pretty good health as yet. There are a great many of the *Dee's* men in bad health and not likely to recover. At 10 a.m. retired to Divine worship in our cabin. At 10 p.m. the ice split again in different places around us which caused us great anxiety of mind. This had been very hard upon us.

26th. Light winds and clear weather called all hands to bring back two of our boats that went off with the ice from along side. They are one half mile from the ship and the road is very very bad. We heard today that there was a man dead in the *Advice* with scurvy. This is the first case of that kind but we have every reason the believe it will not be the last.

27th. Light winds. The Aurora Borealis shining very bright. Our captain got an observation this morning which makes Lat. 70° 24′. We have driven very little south these 10 days. Winds S.W.

28th. Strong gales and heavy snow. At 5 p.m. saw a light at the *Norfolk*. Our captain called all hands and sent us off with lantrons to see what was the matter. When we came there they told us that they wanted no assistance as yet. The ice was split ahead and astern of them and they were out on the ice looking at the crack. Wind North.

29th. Light winds and clear weather. the ice is turning up all around us and a great motion about the *Norfolk* which has caused them to lower down their boats and launch them away from the ship and put out part of their provisions. There is a great berg bearing S.W. from us and we are afraid that it is aground which is making all this disturbance in the ice. Wind W.N.W.

30th. Strong wind. The ice is still at present and all the holes are covered with new ice. The frost is very strong at present. A person cannot walk the deck without the frost taking hold of his face. Some of our men have masks made for their faces. The *Dee's* men are still very bad in health. Wind N.N.E.

31st. Light winds and clear weather. Frost very strong. The ice is fast around us now and nothing to be seen but a solid plain as far as the eye can reach and how much further we know not. The master of the *Dee* is much out of health at present. Wind N.W.

January 1st, 1837. Light winds and variable. The ice is all solid as yet. This is New Year's Day but our table is covered with no fancy dishes. Some of our men have cooked the tail of the fish which we caught which has lain in the hold since it was caught. We have been eating it for some time past and had a large dish prepared for this morning.

2nd. Calm and clear weather. This day we observed a little difference in our daylight. At 2 p.m. our captain called our harpooners in the cabin to know if they would consent to have one piece of beef taken off the ship's company's allowance. They said they would consult the ship's company in order to ascertain whether they would consent thereto, but when the case was made known to them they would by no means consent to it. They stated that they would not take less until the month of March to see what God in His providence would do for us then. Frost severe today.

3rd. Calm and clear weather. Frost stronger every day. Our daylight getting better every day. There is nothing to be seen from our masthead but a solid plain of ice and a number of large bergs. Our master had an observation this morning from a star which makes our Lat. 70° 12½′.

4th. Calm and clear weather. Frost stronger every day, the ice is solid as yet. The *Dee's* men is still getting worse.

5th. Strong gales and heavy snow drift. This is Thursday and our allowance of bread for the week is almost done and we will get no more until Monday. This is old Christmas Day which makes it the more dreary to us when we think on the enjoyments of home and God knows if ever we see such pleasure again, but thank God there is yet hope. There are some of our men complaining with different disorders but to no great extent yet thank God.

6th. Strong winds and dull weather. There is a large lane of water broken out between us and the *Norfolk*. We think we are in a loose piece of ice because our ship turns round 3 or 4 points by compass but all is fast around us as far as we can see except the lane between us and the *Norfolk*. Wind N.W.

7th. Light winds and clear weather. The lane between us and the *Norfolk* is covered with new made ice but has split close to the *Dee's* stern which has caused them all to get ready. Wind E.N.E.

8th. Light winds and dull weather. At 10 a.m. retired to Divine service. The master of the *DEE* is very poorly in health and his crew is always getting worse and some of our men are also complaining very much for weakness and pains in their limbs. Wind W.

9th. Strong wind. People employed in collecting all the spare wood in our ship for firing. We still have a little coals but are afraid all will be little if we are detained in this dreary region. Wind S.

10th. Strong wind. Our master visited the *Dee* and found that Captain Gamblin was very unwell and a great many of his crew in the same state. Several cracks to be seen and the barber flying very brisk on the lanes of water. The frost has been very strong since this month came in. Wind North.

11th. Strong gales and thick snow drift. We are

employed every day in scraping the frost rime off the inside of the vessel. We will get from 5 to 8 buckets full every 24 hours. It comes from the strength of the frost on the dry wood and we have our beds to scrape and clean once a week. Our beds are in a very bad state both wet and cold and dirty. Several of our crew out of health. Frost more moderate today. Wind S.W.

12th. Strong gales and heavy snow. Two of our men travelled to the *Norfolk* and called at the *Dee* as they came back and were informed that one of their men had paid the debt of nature, William Carvigall [Corrigall], an Orkney man, and a number of their men in a dangerous state. Wind E. by N.

13th. Strong wind and dull weather. We hope soon to have the sun by the appearance of the sky. Frost strong. Wind E.N.E.

14th. Fresh breezes and dull weather. Heard a great noise among the ice not very far from us, but it is so dark that we cannot see. The *Dee* men going about their ship with a light. At daylight we saw a great disturbance in the ice but not so near us as we imagined. Our master got an observation today from a star which makes our observation to be in Lat. 69° 53′. Wind N.E.

15th. Strong wind and thick snow drift. Our master told us that the sun was 75 miles from us yet. If our ship continues to drive South we shall we hope soon see the sun as he comes 13 miles North every day. Frost very severe in consequence of which some of our men got frost bit when taking snow on board to dissolve into water. Wind N.E. by N.

16th. Strong breezes and clear weather. At 8 a.m. our master called all hands to launch our boats to the ship and hoist them up. They have been upon the ice since the 22nd of December. This day the crack astern of us opened about 50 yards board. At 8 a.m. it closed and made a dreadful noise. It was very near us. Wind E.N.E.

17th. Moderate winds and dull weather. Our boatswain visited the *Norfolk* last night and they informed him that the crack came to their bows in a very rapid manner and compelled all hands to turn out and land their provisions on the ice, expecting their ship would be lost, but she is yet safe. Thank God for His mercies to us all for we are in such a condition as no man can describe.

18th. Strong winds and clear weather. We have had the comfort this day to see the sun. Thank God for his mercies to us all, for we are in such a condition makes us very much depressed in spirit, but the cheering rays of the sun has again illuminated this dreary region again. All hands that were able to move have been on deck to behold this great renovater of nature, this glorious sight. It is 66 days since we last beheld his cheering beams, but from the appearance among us some will e'er long have to appear before the sun of righteousness to receive a just recompense of reward.[9] Wind N.E. by E.

19th. Light winds and clear weather. Close reefed our top sails and stowed them again. This was done in case of our driving out at an unexpected time and if blowing hard we would not be able to shorten sail being much reduced. This afternoon saw the *Dee* bury another man. Wind N.N.E.

20th. Calm and dull weather. This morning our doctor visited the *Dee* and *Norfolk* to see the sick. He thinks that there are some that will not be long in this world. The ice is very much broken as far as we can see but our ship is fast in the middle of a heavy piece. The days are lengthening fast as the sun is advancing which is a great consolation to us.

21st. Calm and a little snow falling. Our master went to the *Dee* to see how Captain Gamblin was, found him to be very bad and deranged in mind and a great many of his crew in a poor condition. Several of our ship's company complaining very much.

22nd. Sunday. Light winds and dull weather. Retired to Divine service. Our captain had an observation from a star which makes our Lat. 68° 58′. Wind W.

23rd. Light winds and variable. Frost very strong. Our ship in the same position as before.

24th. Strong wind and clear weather. We now have four hours sunlight. It is only six days since we first saw him. Our master took an observation from the sun which makes our Lat. 68° 50′. Wind N.E. by N.

25th. Light winds and clear weather. This day we were

surrounded by several cracks. Some of our men were at the masthead looking out the day being clear but could see nothing but ice and lofty bergs which we have driven past in the darkness of winter. Our crew is getting more sickly and weaker every day. Wind N.W.

26th. Light winds and clear weather. This day we were visited by the master of the *Norfolk*. He told us that his crew was much out of health. We have not heard from the *Advice* since the first of this month.

27th. Light wind and dull weather. Saw several lanes today. The ice is still fast round us. We are driving very little south. Lat. 68° 44′. Wind N.W.

28th. Light winds and variable. The *Dee* buried one of their crew again named Andrew Bennet belonging to Scotland.

29th. Sunday. Light winds and clear weather. At 10 a.m. retired to Divine service. There are several bergs come in sight bearing S.W. of us. We have driven only one mile south since yesterday. Lat. 68° 35′.

30th. Strong gales and a heavy fall of snow. Frost moderate. Wind S.W.

31st. Light wind and clear weather. This day we caught a shark in a hole in the ice on a hook and bait he weighed 250 lb. Our master divided it among the ship's company this we thought a fine fresh mess but was at a loss for fire to cook it with. We were visited by two men from the *Advice*. They told us that their men were all very sickly and their ship making a little water. Wind S.W.

Feb. 1st. Strong gales attended with snow. At 1 p.m. we had a sudden change of wind from the N.E. to S.W. blowing very heavy and a thick snow drift.

2nd. Strong breezes and dull weather. The gale we had yesterday has driven us 5 miles North. Lat. 68° 40′.

At 1 p.m. the *Dee* buried their boatswain. Wind N.

3rd. Light winds and variable. This morning we got another small shark in the same hole weighing 160 lbs. At 9 p.m. there came three men from the *Dee* and told us that their Captain had departed this life. Our Captain took three men and went to the *Dee* and took the body of

Captain Gamblin and put it in the stern boat. The *Dee's* crew is very poorly in health at present.

4th. Fresh breezes and dull weather. Our carpenter and his mate went to the *Dee* to make a coffin for the captain. Lat. 68° 32′.

5th. Light winds and thick snow. At 10 a.m. we retired to Divine worship. At 2 p.m. some of us went to the *Dee* to assist them to bury George Dawson. Wind E.N.E.

6th. Strong winds and dull weather. Our Captain [Captain Taylor] has been at the *Dee* and put the body of Captain Gamblin in the coffin and placed it in the stern boat. Lat. 68° 26′. Wind N.E.

7th. Strong breezes and dull weather. Several lanes of water in sight and several unicorns to be seen on them. Wind N.E.

8th. Light winds and dull weather. At 10 a.m. retired to Divine service. The master of the *Dee* is very poorly in health and his crew is always getting worse and some of our men are also complaining very much for weakness and pains in their limbs. Wind W.

10th. Strong gales and clear weather. This morning we heard that there was another man dead in the *Dee*. We saw a whale in a hole in the ice. Our ship has driven 20 miles South these last two days. Lat. 67° 50′. Wind N.E.

11th. Light winds and dull weather. This day the *Dee* buried another man William Tait belonging to Orkney. We have heard that there were two more of their men dead. Lat. 67° 47′. Wind E.N.E.

12th. Fresh breezes and clear weather. At 10 a.m. we retired to worship in our cabin. This day three men died in the *Dee*. They buried them this afternoon. This was an affecting sight to us to behold. 11 men have died of the *Dee's* crew at this date.

13th. This day we thought we saw Cape Swel but are not certain whether it was land or not. Lat. 67° 29′.

14th. Strong gales. This day we made certain that it was land that we saw yesterday bearing W.N.W. distance 50 miles. Wind N.E. by N.

15th. Strong gales and clear weather. We have driven 18 miles since yesterday. Lat. 66° 59′. Wind N.E. by N.

16th. Fresh breezes and clear weather. This day we saw Dyer's Cape[10] quite plain bearing N.N.W. Several lanes and holes of water to be seen but all the ships remain fast as before. Frost very severe. Wind N.E. by N.

17th. Fresh breezes and clear weather. There is a number of men out of health in our vessel. We have driven 11 miles South since yesterday which makes our Lat. 66° 30′. Wind N.E.

18th. Fresh breezes and clear weather. The cold has been more severe since this month came in than all that we have suffered this winter. We have driven 9 miles South since yesterday which makes our Latitude 66° 21′. Wind N.E.

19th. Light winds and clear weather. At 4 p.m. we retired to Divine worship. Lat. 66° 12′. Wind N.N.E.

20th. Light winds and variable. The ice has broken up, in all directions as far as we can see from our mast head. We have driven very little south since yesterday.

21st. Strong breezes and clear weather. Lat. 66° 6′. Wind W.N.W.

22nd. Light wind and dull weather. Our ship's company is losing their health fast and we are in a very poor state at present. Lat. 66°. Wind N.W.

23rd. Light wind and cloudy. We close reefed our fore sails this afternoon. We have all our sails reefed while we have some strength left for it is daily decreasing and a number of our men are bedfast.

24th. Strong gales and a thick snow drift. Wind N.E.

25th. Strong gales and clear weather. We have driven 30 miles South since last observation. Lat. 65° 20′. Wind N.E.

26th. Strong winds and clear weather. There are two dead bodies in the *Dee* that cannot be buried without assistance from us and the weather is so cold that we cannot travel between the ships. Lat. 65° 4′. Wind N.E.

27th. Light winds and variable. Assisted the *Dee* to bury their dead men. Lat. 65° 4′.

28th. Fresh breezes. At 9 p.m. we saw the *Dee* hoist a light; sent three men to see what they wanted when they told us upon their return that the ice had cracked along side of their vessel and they were getting their provisions in readiness. There are 36 of their men confined to bed and can render themselves no assistance. Wind N.E.

March 1st. A heavy storm and thick snow drift. Several lanes of water in sight. The ship still remaining fast as yet. Wind N.E.

2nd. Light winds and attended with snow. At 1 a.m. the ice broke up with a heavy swell from S.S.E. At 8 a.m. all hands were called to get our ship in readiness for sea. This was cheerful tidings to us all. The mate of the *Dee* came and asked for six of our men to assist them when the ship got into water but Captain Taylor told him he could not be answerable to do so as there were many of our men in a low condition. Our ship has driven 39 miles these last two days. Wind S.S.E.

3rd. Light winds and dull weather. At 7 a.m. we made an attempt to try to clear off a piece of ice that was fast to the bows of the ship. The ice is much broken up around us but we cannot get rid of this piece. We have trenched six feet deep round our bows and have tried to blow it off with powder but all in vain. We then tried to saw it off but to no purpose. At 3 p.m. set the watch and pumped the ship. At 5 p.m. one of our men died, one John West belonging to the wrecked ship. Wind W.

4th. Strong wind from the N.E. and a thick snow drift. The swell is over again and the ice all congealing together again and we are still fast. The frost is so strong that the ice sets together as soon as the swell is over. Wind N.E.

5th. Strong wind and dull weather. Saw the *Dee* burying another man. Our ship has driven 27 miles South these last two days. Lat. 63° 30′. Wind E. by N.

6th. Light winds and clear weather. This day we went to our master and asked him to let us have a little more bread and he allowed us ½ lb. of additional bread per week to each man together with ½ lb. beef to each watch per day in addition to the former allowance. At 9 p.m. George Gilles of the wrecked ship died. Wind variable.

7th. Light wind and clear weather. Several lanes of water to be seen but the ships are all fast. We are longing much for a swell to break up the ice but we hope that it come in moderation for the vessels are in great danger at such a time. Wind N.E.

8th. Light winds and clear weather. We have driven 12 miles South since yesterday. Lat. 63° 6′. Wind N.N.W.

9th. Light winds and dull weather. At 2 p.m. George Flett an Orkney man died. At 4 p.m. we buried him. We had a great deal of trouble before we could get a hole cut through the ice to receive the body. Wind N.E. by E.

10th. Strong gales and thick snow drift. All our men have some trouble on them, some more some less, and some of them in a very dangerous state. Frost severe. Wind N.W.

11th. Fresh breezes and clear weather. Several lanes in sight. Saw the *Advice* with part of her sails set. Saw the *Dee* burying another man making 22 of the crew that have died. Lat. 62° 42′. Wind N.N.W.

12th. Light wind and dull weather. At 10 a.m. we retired to Divine service. The ice is very open to the Southward. Wind N.E.

13th. Moderate wind and dull weather. The ice remaining very open to the southward. Lat. 62° 26′. Wind N.W.

14th. Strong gales from the S.S.E. and thick snow. We can discern a little swell but not so much as broke the ice. We are afraid if it breaks while the storm lasts there will be little hope of our ever seeing our native land. Our Captain has told us the swell will not set in till the wind abates.

15th. Still blowing heavily. At 2 p.m. the swell came on and soon broke the ice up in small pieces. The wind is still from the S.S.E. and a thick snow drift. At 8 a.m. called all hands to saw that piece of ice that was fast to our starboard bow. We kept sawing until 11 a.m. but to no purpose. We then got a sudden change of wind from the N.W. blowing heavily and thick snow, we were then called up off the ice to make sail. At 2 p.m. shipped our rudder, took in four boats. The ice is quite open but we cannot get clear of that piece attached to our bow.

16th. Fresh breezes from the west. At 6 a.m. we got in the open sea after being 5 months and nine days beset. We got out in Lat. 62° 30′ N. We have driven from Lat. 73°

15′ in a continent of ice conducted by the unseen hand of a merciful God. At 8 a.m. called all hands and close reefed our fore and main topsails.[11] At 9 a.m. spoke to the *Norfolk* they told us that they buried three men yesterday. The *Dee* in sight. At 5 p.m. called all hands up that were able to work which were only twenty out of 62 and some of them very unfit for work. We then set the watch thankful to God for this relief. We hope in the course of a month to see our native land once more.

17th. Light wind. At 3 a.m. shook a reef out of our topsails. At 7 a.m. set our main-stay-sail jib. At 11 p.m. set our mizen, people variously employed. Wind N.E.

18th. Wind S.S.E. Standing eastward. At 4 p.m. tacked ship and close reefed our top sails. At 8 p.m. George Gilles of the wrecked ship died. We are in a very poor condition at present. Only sixteen of our crew able to go on deck and consequently very unable to work the ship and to attend the sick.

19th. Wind S.S.E. blowing strong with rain at times. At 6 a.m. tacked ship to the eastward. Took in two reefs in our mizen set our main topsail. No ships in sight.

20th. Strong gales. At 2 a.m. hauled down our main staysail. At 1 p.m. wore ship, close reefed our main topsail. Stowed our trisail. Wind S.S.E. blowing strongly.

21st. Dull weather standing to the Westward. At 2 p.m. James Garson an Orkney man died. At 1 p.m. tacked ship. Wind S.S.E.

22nd. At 2 p.m. got a fair wind. Shook out our reefs and set our main staysail. At 8 p.m. reefed our topsails. Blowing strong. Wind N.E.

23rd. Fine breeze and dull frosty weather. Several of our men in a dangerous state. At 6 a.m. set our main sail. At 8 a.m. William Pain died of scurvy. At 11 p.m. William Stevenson died of scurvy both of them belonging to Shields.

24th. Strong breezes. Ship under close reefed topsails and reefed foresail. At 2 p.m. set the jib. Crew in a very weak state only 15 hands able to go on deck and but few of the Fifteen able to go aloft.

25th. Light breezes and variable with snow. At 4 a.m.

tacked ship. At 8 a.m. sent up our top gallant yard. Wind S.E. by S.

26th Light breezes and variable. At 4 p.m. tacked ship. Wind blowing E.S.E. blowing strong.

26th. Fresh breezes and dull weather. At 5 a.m. strong wind E.S.E. At 8 a.m. Thomas Hunter of the wrecked ship an Orkney man died. At 6 p.m. close reefed our topsails and reefed our mizen and stowed our jib. At 11 p.m. hauled up our main sail.

28th. Strong gales and dull weather. At 2 p.m. wore ship to the Southward. Crew in a very poor state failing fast every day. Lat. 56° 26′.

29th. Strong breezes and dull weather. Ship under close reefed topsails and reefed foresail. At 7 p.m. Robert Elliot died of scurvy.

30th. Strong breezes and cloudy weather. At 9 a.m. set the foresail. At 4 p.m. Andrew Muir died of scurvy. At 8 p.m. increasing wind from the N.E.

31st. A heavy storm and thick sleet from N.E. by N. At 2 a.m. split our mainsail, called all hands, hoisted up our foresail. Split our fore and main top sails, clued them up, none could go aloft. All our sails hanging loose and consequently exposed to the gale. A sea struck us forward and carried away part of our bulwarks. Our sails are all blowing to pieces. We are now in a melancholy state and but a dark outlook, our strength fast decaying and our sails going to rags. At daylight we cut the remaining pieces of our mainsail from the yards and let it go overboard and secured our top sails as well as possible. Wind N.E. by N.

April 1st. Moderate breezes. We are employed in mending our sails. Four of our best hands are taken to bed. At 2 p.m. set our fore top sail. Blowing strong. At 10 p.m. hauled down our main staysail. At 12 o'clock stowed our fore topsail and brought our ship to under a close reefed main topsail.

2nd. Strong gales and dull weather. Crew in a very weak state and few to assist the sick. Wind E. by N.

3rd. Strong breezes with heavy showers of rain. At 8 a.m. wore ship. Set our fore topsail, main staysail and trisail. The few hands that are able variously employed. Wind E. by N.

4th. Moderate wind. At 4 a.m. wore ship. At 7 a.m. set our foresail.

5th. Calm and thick weather with a heavy swell from E.N.E. Only 8 hands able to keep watch.

6th. Light breezes and variable with thick fog. At 4 p.m. a heavy rain. At 8 p.m. strong gales from W.N.W. with rain.

7th. Strong gales from the west and dull weather. At 4 p.m. Peter Johnston died of scurvy.

8th. Moderate breezes and dull weather. At 3 a.m. James Harsburs died of scurvy.

9th. Fine breezes with showers. At 10 a.m. shook one reef out of our main topsail. Crew getting weaker.

10th. Light breezes and dull weather. At 2 p.m. John Fraser second mate of the wrecked ship died of scurvy. This is the thirteenth of our crew that has died and by all appearance there soon will be more. There is not one free of disease and the greater number of us in a very dangerous state.

11th. Light breezes from the West and dull weather. Set our jib and mainsail. At 2 p.m. a foul wind.

12th. Strong breezes throughout with thick weather. The men who are able to keep watch are eight in number most of them very weak and those who are confined to their beds getting weaker every day.

13th. Moderate breezes and clear weather. Lat. 59° 25′.

14th. Moderate breezes and clear weather. At 10 a.m. Thomas Wilson of the wrecked ship died of scurvy.

15th. Heavy squalls attended with snow and hail. At 1 p.m. Antony Bruff died of scurvy.

16th. Moderate winds and clear weather. We are all in poor condition of health, most of our crew despairing of life. Wind still foul. Only six able to go on deck and none able to go aloft. We have a stool at the wheel to sit on and steer and a pot of cold water placed along side to keep us from fainting.

17th. Heavy squalls from the West and a very heavy sea. The ship under close reefed top sails and reefed fore sail. We are not able to make sail although the wind is fair. At 2 p.m. Robert Littlejohn our carpenter died. He was a few days ago one of our best hands but being hove over the wheel was knocked almost dead on the quarter deck. He never recovered his reason and remained in that state until he died.

18th. Moderate winds and clear weather. Scarce any able to go on deck. This day we are in such a condition as is beyond our power to describe. At 8 a.m. Joseph Hoggard our cooper died of scurvy. At 3 p.m. Magnus Fiddler died of scurvy. At 6 p.m. Hugh Seater died of scurvy. We are scarcely able to get their bodies overboard.

19th. Moderate winds and clear weather. Crew in a very weak state almost despairing of life. Wind right in head of us this morning. Ship ratching to the North. The same sail set as before. At 1 p.m. the second mate went up to the mast head to see if he could see any vessel. He got as far up as to the mainstay[12] when he spied a ship to windward, this was glad news to us. We immediately hoisted our signal of distress and he came running down upon us until he came within hail of us and then run away. She was a barque belonging to Hull. She was not far from us until we spied another bearing down upon us. We saw two more and before dark there were seven sail hove-to around us. The first one that came to us was a brig belonging to Hull. The Captains of the seven vessels as soon as they were informed of our miserable condition agreed to give us a man from each of their vessels to assist us to work the vessel. This was such a comfort to us as we are unable to express. They also gave us a supply of what provisions they could spare such as they judged most suitable for us in our sad state. Those men who came on board of us were very kind and rendered every assistance to us that was in their power not only in working the vessel but also on supplying us with the necessities of life.

20th. Those men sent to our assistance sent up our top gallant yards and made all sail on the ship that could be allowed. There were no more deaths on board until the 24th when Thomas Care left this world. He was the last one that was rolled up in his blanket and sunk in the ocean. The man that came first to us brought along with him some fresh beef and some barley, cabbage, potatoes, coffee, sugar and bread and told us that he was glad he had fallen in with us. He came down in our half deck and went to every bed cabin where there was any person sick. There was nothing more in particular that happened until the 27th being the one on which we arrived in Stromness Harbour.

At 11 a.m. the doctor appointed by the Government came on board and got the sick conveyed on shore to the hospital. They are all recovered with the exception of William Learmonth and John Wilson who died in the hospital. The state in which our ship was before we came to Stromness cannot be described.

The manner in which we buried our dead at sea was as follows. Each body was rolled up in a blanket, iron bolt put to the feet and a board so placed as to slip off the body from the deck into the ocean.

It may be stated that Captain Taylor was a man of exemplary piety, did all in his power for the comfort of his men both for time and eternity, but it is much to be lamented that in general they were very unwilling to be instructed by him until it was too late which was confessed by some of them before death.

150 years on . . .

The Story of the Orkney Natural History Society and Museum.

The Early Years

Like many country people, Orcadians have always shown great interest and knowledge in their history and in their environment, because this same environment has a considerable influence on their daily lives.

The 19th century in Scotland was a time of great scientific enquiry, and it seems that in Orkney a need was felt for a society that could devote itself to the natural history of the county. It appears that considerable interest was shown in the founding of such a society, and, accordingly, on 28th December 1837, a public meeting was called in Stromness. In the chair was the Rev. Dr Charles Clouston of Sandwick, a notable amateur scientist who would hold the position of President of the Orkney Natural History Society for 48 years.

The Meeting agreed a draft constitution, of which the main points are as follows:

1. The Society shall be called the Orkney Natural History Society, and shall have as its object the promoting of Natural Science by the support of a museum and by any other means in its power.
2. While special care shall be used to collect specimens for the museum of such objects in Natural History and Antiquities as Orkney can furnish, the Society shall use all proper means to enrich its collections with specimens in Natural History and Antiquities from any other quarter.

The usual arrangements were made for the election, as

Stromness 28th December 1837

At a general Meeting of the Subscribers to the Prospectus of a Natural History Society for the County of Orkney, Revd. Mr Clouston of Sandwick in the chair

The meeting agreed to Constitute itself into a Society called the Orkney Natural History Society, when the following regulations which had been prepared were now read, considered & unanimously approved of —

1. This Society shall be called the Orkney Natural History Society & shall have for its object the promoting of Natural Science by the support of a Museum, & by any other means in its power.

2. While special care shall be used to collect for the Museum specimens of such objects in Natural History & Antiquities as Orkney can furnish, the Society shall use all proper means to enrich its collection with specimens in Nat. History & Antiquities from any other quarter.

3. The property in the Museum shall be

Minute of the inaugural meeting, 28th December 1837.

Honorary Members, of "Gentlemen of reputation for Science or who have contributed handsomely to the Institution." We do not know how many attended this inaugural meeting, but the meeting of 29th July 1838 gives a nominal roll of a hundred and five members, from all over Orkney and beyond. The annual subscription was fixed at Two Shillings. In the early days, admission to the Museum Room was limited to committee members only, but later it was extended to include members and friends. Other visitors were required to pay sixpence each.

Throughout the 19th and early 20th centuries, members were active in natural history and antiquarian studies, producing papers which were read to the Society and published in the local press. These included such titles as "Hints on the Geology of Orkney" by the Rev. John Gerard of South Ronaldsay, "The Ruins of Breckness: Prehistoric and Modern" by W. G. T. Watt of Skaill, and "The Habits of Birds Frequenting Sule Skerry" by James Tomison. A selection were published by the Society in 1905 in a volume entitled *Orcadian Papers.*

To house its exhibits, the Society rented, for two guineas a year, "Mrs Flett's large room", situated on an upper floor of Flett's Commercial Hotel, at the corner of Church Road and Victoria Street in Stromness. A woman was appointed to keep the room clean, her duties being strictly laid down, but later a man, James Sinclair, was appointed as Curator, at the salary of £3 per annum. When Captain Flett, the owner of the premises, gave notice that he was no longer prepared to renew the lease, the Society decided to acquire its own premises. At the same time the Stromness Town Council was also looking for premises, so the two bodies combined to erect the present building, opened in 1858, with the Museum upstairs and the Town Hall below. In 1929 the Town Council found new premises, and the Society took over the whole building, thus doubling the exhibition space. The association between the Town Council and the Orkney Natural History Society was a happy one, though in the initial stages both parties were hampered by shortage of money, and it was some time before the building was made fully fit for its purpose.

The Collections

From its very beginning the Society received many donations of specimens of a very wide-ranging character. The early Minutes record such items as:

Model of an Eskimo sledge and dogs.
Marble from the Parthenon at Athens.
Gun screw from the Battle of Waterloo.
Wooden door locks from North Ronaldsay.
Dyewood from the wreck of an East Indiaman in North Ronaldsay.
Four silver leaves from Napoleon's grave in St Helena.

In 1840 Captain Taylor of the whaler *Grenville Bay* donated "two bears' heads, plumbago, and coal specimen from Davis' Straits."

In its early years the Society included among its members two eminent Victorians — Hugh Miller, the celebrated geologist and author of *Footsteps of the Creator,* who donated his important fossil, *Homosteus milleri,* to the Museum, and John Rae, the Orcadian Arctic explorer who discovered the fate of the Franklin Expedition. Their memorabilia are included in the Museum collections.

The exhibition of birds, occupying the upper floor of the Museum, has been built up over the years from many donations, and is considered to be a particularly fine one.

The first mention of the intention to form a collection of birds occurs in the Minutes of the Society in October 1839, when it was agreed ". . . to send to a proper bird-stuffer in the South what rare Orkney birds might be obtained, and also to make a vigorous effort to build up the Ornithological Dept. of the Museum with as much despatch as practicable."

Various collections of eggs have been received, forming an almost complete collection of the eggs of Orkney breeding birds. Other valuable Natural History collections in the care of the Museum are the Magnus Spence Herbarium of Orkney Plants, the Robert Rendall Collection of Orkney Shells and Seaweeds, and the Lorimer Collection of Orkney Butterflies and Moths.

REV. JAMES RITCHIE,
PRESIDENT, 1885-97.
REV. WILLIAM STOBBS,
SECRETARY, 1837-63.
DR. CHAS. CLOUSTON,
PRESIDENT, 1837-85.
MR. WILLIAM ROSS,
SECRETARY, 1863 – 70.
REV. J. S. NISBET
SECRETARY, 1870-74.

The Museum above the Town Hall in the late 19th century.

Hand harpoons from the Kirkwall whaler *Ellen,* along with a flensing knife for cutting blubber.

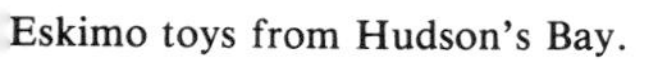

Eskimo toys from Hudson's Bay.

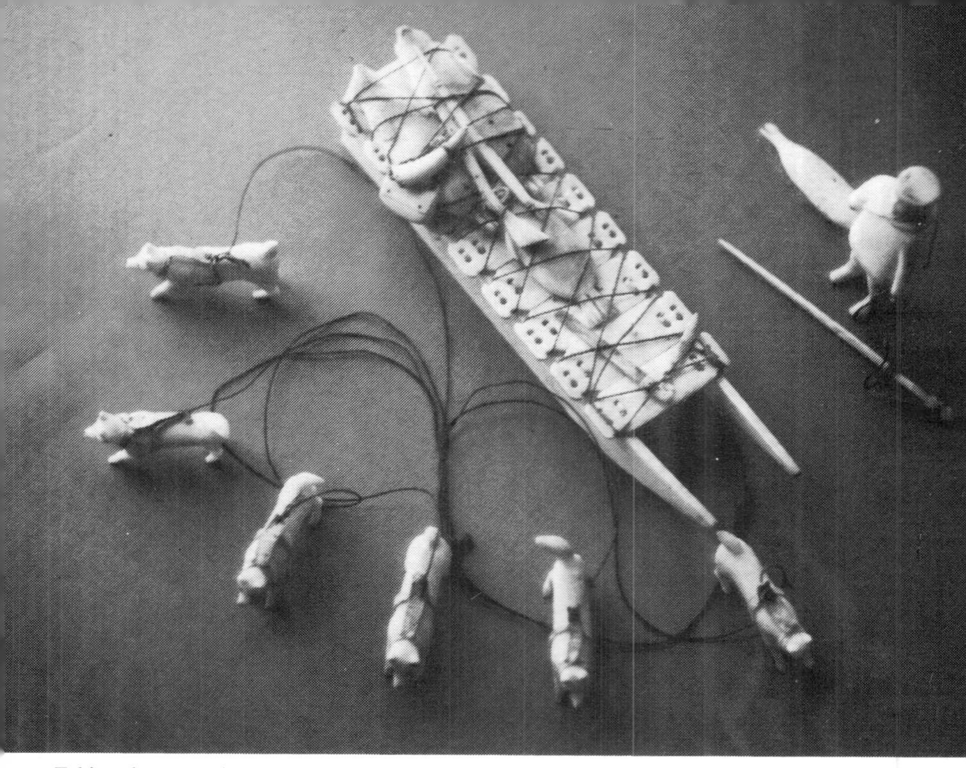

Eskimo bone carvings.

Powder horn from the Franklin expedition.

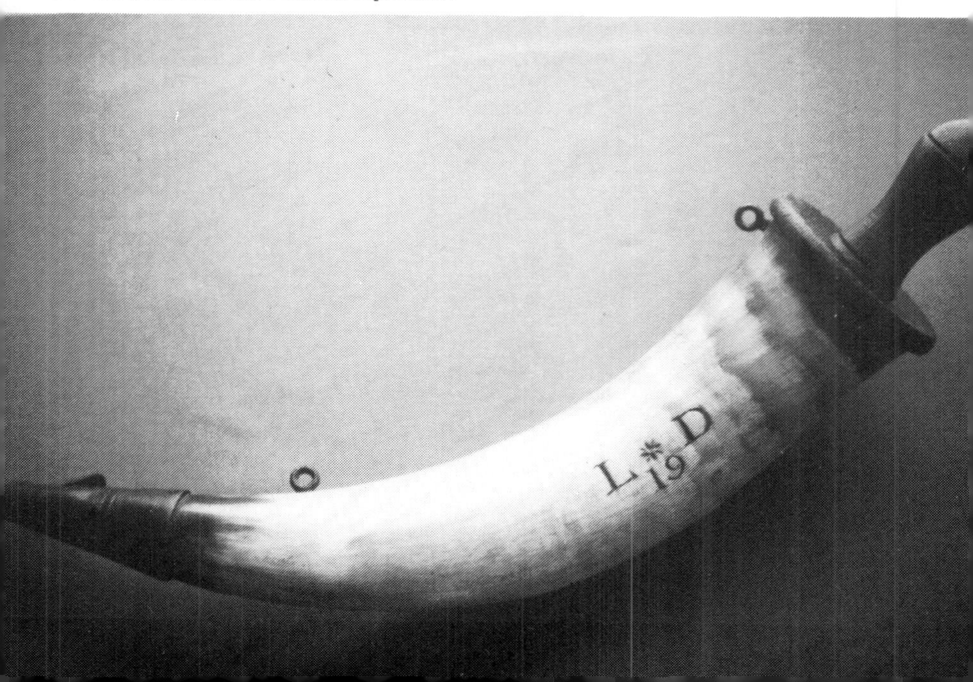

The Hudson's Bay Company flag.

A Greenland Falcon in the bird collection.

Homosteus milleri.

The present custodian, Mrs J. J. M. Firth, a veteran of 25 years' service, at her sales counter.

Museum custodian 'Mammy Lyon', photographed in 1912.

Collecting Policy

For many years the Society had sole responsibility for collecting in all aspects of Orkney history and prehistory, but this is no longer the case. The Society now sees its role as portraying the Natural and Maritime History of the islands — the areas in which its collections are strongest — complementing the archaeological displays at Tankerness House Museum in Kirkwall and the Orkney Farm Museum in Birsay and Harray. In this it has had the advice and practical help of Orkney Museums Service. By concentrating on fewer themes, these can be more deeply researched and better displayed.

The maritime history of Orkney is particularly rich. The Society has been fortunate over the years in being able to collect material on the Hudson's Bay Company, whose ships called at Stromness for provisions and workers until the end of the 19th century. Stromness was also a regular port of recruitment for the Arctic whaling fleet, and one of the main Orkney centres of the 19th century herring fishing. Within recent years the Museum has built up a display on the German High Seas Fleet, scuttled in Scapa Flow in 1919, which is currently of great interest to sub-aqua divers.

Over the years, the Society has built up an excellent archive of local history photographs. To the glass negatives of George Ellison and Willie Hourston have been added, through the good offices of the Orkney Library, collections of prints by R. H. Robertson and Tom Kent, giving a detailed and intimate view of Stromness in the late 19th and early 20th centuries, and the various aspects of Orkney maritime history. Two collections of photographs, including a fascinating journey through Stromness by an anonymous Victorian photographer, have been published.

Special Exhibitions

Museum displays were static until 1966, when the Society began a regular programme of Summer Exhibitions on local history themes, such as "The Lighthouses of Orkney", "For Those in Peril" (Orkney Lifeboat history) and "Harvest of Silver" (the herring industry in Orkney). As well as greatly increasing visitor numbers, these have

Honorary President and Curator James G. Marwick, tailor and naturalist.

had the effect of systematically increasing the Museum's collections and store of knowledge.

Publications

Exhibitions have frequently been accompanied by informative booklets, adding the dimension of publishing to the Society's activities. A recent venture, in association with the Orkney Press, has been a guide to the wrecks of Scapa Flow. *Reminiscences of an Orkney Parish,* first republished by the Society in 1974, has been a perennial best-seller.

The Future

The Orkney Natural History Society has come a long way in the past 150 years — its willingness to adapt to new circumstances has ensured its survival where others have succumbed. The emergence of the Orkney Field Club and the Orkney Heritage Society, not to mention the activities of the Royal Society for the Protection of Birds and the Nature Conservancy Council, have allowed it to concentrate on the absorbing task of maintaining and developing its Museum. In this it has the close co-operation of the Orkney Islands Council, with whom it looks forward to a long and fruitful partnership into the 21st century.

Membership of the Orkney Natural History Society, a registered charity, is open to all — funds go directly towards maintaining the building and improving the care and display of exhibits. Current rates can be had on application to the Museum.

Peter Leith and David Anderson,
Orkney Natural History Society and Museum,
52 Alfred Street,
Stromness,
Orkney.

Honorary Presidents of the Orkney Natural History Society

Rev. Charles Clouston	1837-1885
Rev. James Ritchie	1885-1897
Mr J. D. Turner	1897-1899
Mr M. M. Charleston	1899-1905
Dr Grant	1905-1908
Mr W. McKay	1908-1927
Col. H. H. Johnston	1927-1931
Mr J. G. Marwick	1931-1935
Mr J. C. Kelly	1935-1939
Mr J. W. Towers	1939-1944
Mr J. E. P. Robertson	1944-1965
Dr J. Cromarty	1965-1971
Mr W. Groundwater	1971-1976
Mr P. K. I. Leith	1976-

Publications

Exhibition booklets—
The Orkney Croft.
Sail and Steam.
The Salving of the German Fleet.
The Lighthouses of Orkney.
Harvest of Silver (the herring fishing in Orkney).
For Those in Peril (Orkney Lifeboat history).

Stromness — late 19th Century Photographs.
Reminiscences of an Orkney Parish by John Firth.
Stromness Before the Great War — photographs by George Ellison *(now out of print).*
The Wrecks of Scapa Flow (in association with the Orkney Press).

Exhibitions

1966 Orkney Airmail Service.
1967 The 150th Anniversary of the Burgh of Stromness.
1968 Robert Rendall Commemorative Exhibition.
1969 Orkney Postcards.
1970 The 300th Anniversary of the Hudson's Bay Company.
1971 Old Stromness and Round About.
1972 The Orkney Croft.
1973 Sail and Steam.
1974 The Salving of the German Fleet.
1975 The Lighthouses of Orkney.
1976 Harvest of Silver.
1977 For Those in Peril.
1978 The German Fleet in Scapa Flow.
1979 Old Stromness — Photographs by R. H. Robertson.
1980 Stromness — Photographs by Tom Kent.
1981 Willie Hourston, Photographer.
1982 Days of Cord and Canvas.
1983 The Sunken Fleet in Scapa Flow.
1984 The Wreck of the *Svecia*.
1985 The Sinking of the *Royal Oak*.
1986 Eliza Fraser, Castaway.
1987 The Ice-bound Whalers.

THE ORKNEY PRESS

The Orkney Press was set up in 1981 as a non-profit-making organisation to provide finance and professional skills for local publications. Titles are as follows:—

Willick o' Pirliebraes by David Sinclair *(now out of print)*
Island Images by Betsy I. Skea
The Men of Ness by Eric Linklater *(republication)*
Orkney Short Stories, introduced by George Mackay Brown
Stenwick Days by R. T. Johnston
The Symbol Stones of Scotland by Anthony Jackson
The Wrecks of Scapa Flow, compiled by David M. Ferguson
Ancient Orkney Melodies by David Balfour *(republication)*
The Voldro's Nest by Margaret Headley

In the ASPECTS OF ORKNEY series:—

1. *Kelp-making in Orkney* by William P. L. Thomson
2. *The Birds of Orkney* by Booth, Cuthbert and Reynolds
3. *This Great Harbour — Scapa Flow* by W. S. Hewison
4. *The People of Orkney,* edited by R. J. Berry and H. N. Firth

NOTES

Orkney and Arctic Whaling

1. Britain 1986: an official handbook, p.309

   Cargo Handled in Million Tonnes

Sullom Voe	60
London	48.2
Forth	29.8
Orkney	16.1
Liverpool	10.8
Clyde	10.5

   For fuller particulars on the Scottish ports see Scottish Abstract of Statistics 1985, p.169, 183-4.
2. Cumbria Record Office, Carlisle D/He/1/55
   Letter dated 27th August 1841 from E. F. Sheppard, a visitor to Stromness. It is interesting that the young naturalist throughout uses the whaleman's inaccurate term 'fish' to describe these mammals.
   Through the good offices of Jeremy Godwin a copy of Sheppard's letters now resides in Stromness Museum.
   Cwt. — hundredweight (112 lbs., 1/20 ton).
   Pre-decimalisation, the pound was divided into 20 shillings (s. or /-), each of which in turn gave 12 pence (d.).
   For a similar whalehunt see *Orkney Book* ed. J. Gunn (1909) pp.242-7 which is taken from *Summers and Winters in the Orkneys* by Daniel Gorrie (London and Kirkwall 2nd ed., n.d.) pp.215-227.
3. *An Account of the Arctic Regions with a History and Description of the Northern Whale Fishery* by William Scoresby (Edinburgh 1820) I p.464 & 474.
4. Lancelott Anderson of Hull quoted in *History of the Whale Fisheries* by J. T. Jenkins (London, 1921) p.149.
5. *Yankee Whalers in the South Seas* by A. B. C. Whipple (London, 1954) p.139.
6. *Old Whaling Days* by W. Barron (Hull, 1895) p.28.
   Flensing is the stripping of blubber from the whale.
7. Whipple p.139.
8. Scoresby I, pp.239-241, 284.
9. ibid. II p.389
   Making off is the entire process of stowing the blubber in barrels and all the cutting that leads up to that.
10. *British Whaling Trade* by Gordon Jackson (London, 1978) pp.84-5.
    Scoresby II chap.vi., *The Arctic Whalers* by Basil Lubbock (Glasgow, 1936) pp.183-4.
11. Jackson p.29.
12. ibid. pp.54-68 for a fine analysis of the economic factors affecting whaling development.

13. *Pen and Pencil Sketches of the Whale and Seal Fisheries in the Arctic Regions* by R. H. Hilliard p.76.
Microfilm in City of Dundee Museums collection, original journal in Glenbow Museum, Calgary, Alberta.
14. *Journal of a Voyage to Baffin Bay* by John Wanless, surgeon of the *Thomas* of Dundee 1834.
Entry for Sat. 28th. June.
15. *A Geographic and Hydrographic Survey of the Orkney and Lewis Islands* by Murdoch Mackenzie (London, 3rd edit. 1776) p.1.
16. Barron p.61.
17. Barron, p.25.
18. *Greenland Voyager* by Tom & Cordelia Stamp (Whitby, 1983) pp.3-4.
See also Scoresby II, pp.187-191.
19. Scoresby II, pp.197-8.
20. *Life on a Greenland Whaler* by A. Conan Doyle in Strand Magazine (London, 1897).
21. Wanless, Mon. 16th. June 1834.
22. Stamp, p.92.
23. *Grenville Bay* is given as 328 tons by Lloyd's Shipping Register, as 306 tons in *History of South Shields* by G. B. Hodgson, p.303, which also records her as built in 1783. (Miss Peterson's notes.)
Dee of 319 tons joined the Aberdeen flett in 1814.
Whale and Seal Fisheries of the Port of Aberdeen by James Pyper, p.44 in The Scottish Naturalist No. 176 (1929).
24. *Robertson Letter Book.* Mrs Christian Robertson to William Anderson, Manager for Whale Fishing Company, Montrose 27/12/1824.
"The best men's wadges are 35/- & 1/-p.T. and as low as 16/-p.m. & 6d. according to the goodness of the men they have got . . . The men are very plentiful here. They will have no trouble in manning the ships." (to G. & J. Egginton, Hull 9 March 1830)
25. Shetland Archives. GD150/2518B, Morton Papers. Memorial for the Rt. Hon. the Earl of Morton on behalf of the Gentlemen, Heritors, Merchants, and other Inhabitants of the Islands of Zetland, 14th. August 1756.
Shetland Life and Trade 1550-1914 by Hance D. Smith (Edinburgh, 1984) p.88.
26. 46 George III c.9 pp97-98.
Jackson pp.72-3, but see also p.63 for effect of press gang.
27. Scoresby II, p.383.
28. Sheppard, Ipswich, 28 Sept'r 1840
"There are but very few species of land birds inhabiting Orkney for with the exception of what parts are cultivated and sown with grain the whole of the Islands are just as naked and bare as English Heath Land with all the furze bushes, brakes &c. cleared off it."
Mrs B. Robertson to G. & J. Egginton, Hull. 10 Aug. 1819
"I only sent 2 oxen by Capt. K. as I could not ship any more without delaying the vessel (it being Sunday)."
29. *The Present State of the Orkney Islands* by James Fea (pronounced Fay) (Edinburgh 1775) p.26.
30. *Narrative of a Voyage to Hudson's Bay* by Lieut. Edward Chappell, R.N. (London 1817, facsimile reprint Toronto 1970) p.14.
31. The Statistical Account of Scotland 1791-1799:
Vol. XIX. Orkney and Shetland (Wakefield 1978).
Evie and Rendall 1797, p.78. Aithsting and Sandsting 1792, p.390.
Kirkwall and St Ola 1791, p.135. Delting 1790, p.412.
S. Ronaldsay and Burray 1793, p.194. Unst 1791, p.514.

St Andrews and Deerness 1797, pp.206 & 208.
Stromness and Sandwick 1794, pp. 237, 249-250.
The most informative on this matter are Delting and Stromness.

32. Lubbock, p.119.
Fishing for the Whale by David S. Henderson (Dundee 1972), p.10.
33. Jackson, p.89, n.7.
34. O.S.A. Stromness and Sandwick, p.254.
35. Fea, pp. 101, 104.
36. ibid. pp.7-8.
37. Disposition of sale of land at the east end of the shore of Kirkwall, 14 June 1813. By courtesy of John D. M. Robertson as are the dispositions n.38. The 'oily house' stood on the site of the office of S. & J. D. Robertson.
Scottish Record Office — bounty papers of the *Ellen* 1813-15, 1817-20, all in general classification E508, e.g. 1813 — E508/116/8/30;
1818 — E508/120/8/50.
The *Ellen* was a vessel of 279 47/94 tons. She carried 5 whaleboats and 39 men.
38. Dispositions, 10 April 1823 & 6 Sept. 1834, the latter taking effect at Martinmas 1829.
Certificate of service of Francis Burnett, 1805-1833. Stromness Musuem collection.
Kirkwall in the Orkneys by B. H. Hossack (Kirkwall, 1900) pp.412-3.
39. Stamp, chap. 6.
40. Barron, p.17.
41. Scoresby document May 3rd. 1823 in Mystic Seaport Museum, Connecticut, quoted by Mrs Cordelia Stamp in letter 3 Sept. 1986.
42. The Statistical Account of The Orkney Islands (Edinburgh and London 1842), p.34.
43. Jackson, pp.129-130.
44. *Orkney Herald* 27 March 1866.
45. Jackson, p.125, table 10.
46. Mrs C. Robertson to G. & J. Egginton 21 Apr. 1825.
47. *North-east Scotland and the Northern Whale Fishing 1752-1893* by R. C. Michie, pp.71-2 in Northern Scotland Vol. 3 no. 1 (Aberdeen 1977-8).
48. Jackson, p.126.
49. *Sketches of Davis' Strait, and and account of the disasters there — season 1830* by James Cumming (Aberdeen Magazine i. 1831) p.74.
50. ibid. p.76.
51. ibid. p.78.
Lubbock pp.278-284 carries a full account of the episode.
52. Anon. The Sufferings of the Ice Bound Whalers (Edinburgh, 1836) p.9.
53. ibid. p.17.
54. W. Elder manuscript journal p.54, quoted in The Drift of the Whaler *Viewforth* in Davis Strait 1835-36 by Alan Cooke and W. Gillies Ross. (Polar Record Vol. 14 no. 92, Cambridge 1969) p.588.
55. The Voyage of *H.M.S. Cove,* Captain James Clark Ross 1835-36 by A. G. E. Jones (Polar Record Vol. 5 no. 40, Cambridge 1950) p.544.
56. John o' Groat Journal & Caithness Monthly Miscellany 1836.
57. Jones p.548.
58. Orkney Archive. Balfour Papers. D2/28/10 John Baikie to Thomas Balfour, M.P., 24/2/1836.
59. ibid. James Login to Thomas Balfour 14/3/36.
60. ibid. Baikie 24/2/36.

61. ibid. Login 14/3/36.
Ireland is a tunship in the parish of Stenness, Birsay, Firth, Rendall and Orphir are all parishes in the West Mainland of Orkney.
See also D2/4/2 W. G. Watt to T. Balfour quoted in Thomas Balfour by Irene Rosie (Kirkwall, 1978) p.9.
62. Orkney Archive. D2/28/10. A Plummer to T. Balfour.
Compare that to James Login's comment about the same ship:— ". . . a woful change came on them caused chiefly from want of fire, having burnt all their loose wood . . . Their clothes gave way which obliged many to take to their beds which are now in a state too shocking to be described."
63. Wanless. Thur. 6 May 1834.
64. Arbuthnot summary lists. Peterhead Museum.
65. Orkney Archive. D2/28/10. Login to Balfour 14/3/36.
66. Lubbock p.54.
67. John o' Groat Journal 12 May 1837.
68. Lubbock pp.337-9.
69. Medicine and the Navy 1200-1900 by C. Lloyd & J. L S. Coulter Vol IV — 1815-1900 (Edinburgh & London, 1963) p.107.
70. ibid. p.115; Lind's Treatise on Scurvy ed. by C. P. Stewart & D. Guthrie (Edinburgh 1953) p.406.
71. Lind pp.222-3, see also p.91.
72. ibid. p.88.
73. Lloyd & Coulter IV pp.108-9; Lind p.161.
74. Lind p.215.
75. Lloyd & Coulter IV p.119.
76. Surgeon, R.N. June 20, 1837 in Dundee, Perth & Cupar Advertiser, but originally appearing in Montrose Review.
77. Aberdeen Chronicle 28 Dec. 1836. Letter of John Barrow to A. Bannerman, M.P., dated 23 Dec.
78. Orkney Archive, D2/20/17 C. Wood, Secretary to the Admiralty to T. Balfour, M.P. 13/1/1837.
79. John o' Groat Journal Dec. 1836.
"The Admiralty will, we trust, make the necessary provision for their reception at Stromness, so that, should our hopes be realised, the crews may not be subjected to still further sufferings upon their arrival in British territories."
80. Aberdeen Chronicle, 11 Jan. 1837. Speech by Councillor Bisset, presenting a motion to Aberdeen Burgh Council to send a petition to London.
81. Dundee, Perth & Cupar Advertiser, Jan. 20th 1837.
"My Lords will be prepared to pay the sum of £300 to each of the first five vessels which may sail from any port in England or Scotland before the 5th. of February, carrying an extra quantity of provisions, provided they show by their log that they make the best of their way across the Atlantic, and that they reach the edge of the ice to the southward of 55 Latitude.
"My Lords are willing . . . to defray twice the value of any Provisions supplied to any of the distressed ships . . . which may be met with on their passage home, and the wages of any men put on board them for the purpose of navigating them home."
Letter from T. Baring on behalf of the Admiralty to the Provost of Dundee.
See Lubbock p.328.
82. John o' Groat, Feb. 1837. Lubbock pp.328-9.
83. Newcastle Chronicle, 13th May, 1837.
84. Newcastle Chronicle, 6th May, 1837.
"We received supplies from the *Lady Jane,* Capt. Leask; the *John and*

Jane, of Hull; the *Ellen,* of Leven; the *Harvest Home* and the *Stephen,* of Newcastle; the *Loyal Briton,* of Whitby; and the *Perseverance,* of North Shields. We had 20 deaths in all, 10 of our own crew and 10 belonging to the *Thomas.* The most of our crew will be left in the hospital here. They began to recover rapidly after receiving the supply of fresh provisions.''

A fuller version of this letter from Capt. Thomas Taylor is in Lubbock p.330.

As to cabbage, Lind noticed in scurvy sufferers, ''the most craving anxiety for green vegetables, and the fresh fruits of the earth.'' Lind p.98.

85. Lloyd & Coulter III (Edinburgh & London, 1961) p.30, quoting 'Memorial Concerning the Present State of Military & Naval Surgery' by John Bell, a leading anatomist. 1800.
86. Lloyd & Coulter IV pp.12-13, 16, 45-46.
    Public Record Office, A.D.M. 104/21/384.
    John Hamilton was probably using his long periods of absence from the navy to build up a practice in Stromness. The 1841 Census shows him as a surgeon in the burgh. During the 1850s he was acting both as surgeon and as Lloyds' agent.
87. Dundee Advertiser, 12 May, 1837.
88. John o' Groat, 26 May, 1837.
89. For fuller accounts of the *Advice* and *Swan,* see ''The Whaleship *Dee* — a sad winter's tale of 150 years ago'' by J. A. Troup in 'Orkney View', Oct./Nov. 1986 and, above all, Lubbock pp.332-341.
90. Orcadian 26/3/1936.
91. Lind p.339, quoting Richard Walter, surgeon on Anson's voyage round the world (1740-4).
92. *The Importance of Ascorbic Acid to Man* by A. P. Meiklejohn, R. Passmore and C. P. Stewart in Lind pp.437-8.
93. Dundee Advertiser 12/5/1837, quoting the log of the *Washington.*
94. I am indebted to the Flett family at Nistaben, to Ronald Shearer and to Peter Leith for information about Adam Flett.

NOTES

Account I — *The Dee*

1. Ice terminology:
   Land ice was the immediate field of ice firmly attached to the coast.
   A floe was a very large piece of ice, at least half a mile in diameter.
   Bay ice was ice that formed as the surface of the sea froze. It could form rapidly. Scoresby indicates an inch over 24 hours of keen frost increasing to a foot of thickness in a month of calm frosty weather. Even as a sludgy material it would quickly halt a sailing ship.
   Sconce pieces were smallish flat sections of shattered floe or new ice that would finally freeze together to form pancake ice.
   Light ice was up to a yard thick; heavy ice was in excess of that depth.
   Scoresby I pp. 225, 239-241

2. The *Friendship* must have left Pond Inlet about this time and have found a route through the ice. Her pay book indicates a voyage of 7 months 5 days starting on 17th March and therefore returning to Dundee on or about 22nd October. Payment at full monthly rate to all crewmen suggests that she suffered no fatalities. All the crew joined the vessel at Dundee.
   The catch of "11 sizeable fish . . . 1 dead fish" was estimated to yield 74 tons of oil. Thus a seaman would earn £12 10s 10d in basic wages and £4 12s 6d in oil money.
   Paybook of the *Friendship* is in the City of Dundee Museum collection in Broughty Castle Museum.

3. Cape Melville at the northern end of Melville Bay, some 75 miles east of Cape York. The Duck Islands lie just south of Melville Bay.

4. Ford Littlejohn was a well-educated young man who had studied Arts in St Andrews University from 1827 to 1832 and Medicine in Edinburgh from 1833 to 1836. In 1837, after the return of the *Dee* he graduated as L.R.C.S.E. (Licentiate of the Royal College of Surgeons, Edinburgh.)
   (Information from Mr R. Smart, Keeper of the Muniments, St Andrews University and from Dr R. Cant.)

5. William Corrigal, "native of Stromness" does not appear in the 1821 Census of Stromness.

6. One of Adam Flett's stories was of throwing blankets out on the ice with the aim of destroying the lice, but when they got to bed " the little beggars were as active as ever."

7. James Tulloch of Harray is sometimes described as dying in sight of Orkney. This death entry shows the fallacy of that. He was taken home by his wife's cousin, Adam Flett, and buried in St Michael's Churchyard. The three week time lag between death and arrival in Stromness explains why Adam Flett, in making the coffin, tarred it inside and out.
   (For the last two foot-notes I am indebted to Mr Bertie Flett of Nistaben, Dounby — great-grandson of Adam Flett.)

NOTES

Account II — *The Grenville Bay*

1. The Devil's Thumb, a dramatic and unmistakable shaft of rock that marked the southern end of Melville Bay.
2. Refraction of light in the Arctic can sometimes give a clear picture of an area beyond normal field of vision.
3. To harpoon the whale it was necessary to row the 18-foot whaleboat close up behind the whale's head. Otherwise, when the heavy tail flukes lashed the water, the boat and men were at risk.
4. On the whaling grounds ordinary land times were used instead of timing by bells to fix watch changes.
5. The whaleman's term, 'milldolling', covered two possible modes of opening a track before the ship in thin ice. Sometimes it was adequate to hang a boat from the jib boom with one or more men aboard to keep it rolling. Alternatively, a boat, preferably old, could be continuously hauled up to the bowsprit and allowed to fall on the ice thus breaking a patch of ice and enabling the ship to keep edging forward.
6. This seems to mean that the boat was retrieved and taken back on deck. Compare similar wording of 16th January.
7. The suggestion has been made that the captain, Thomas Taylor, belonged to Brecks, S. Ronaldsay. That is unlikely, particularly on account of the youth of the S. Ronaldsay man, but not impossible. When the *Lady Jane* of Newcastle was beset by ice in the previous winter Captain Leask was described as "a Stromness man." (Orkney Archive — Balfour Papers D2/28/10). One would expect any Orkney connection to be mentioned by the two E. Mainland men who wrote this text, if they knew of it, or even suspected it.
8. Possibly one of these was Adam Flett's brother, even though the family speak of his crossing the ice to the funeral. This is the only occasion when death **aboard** the *Thomas* is mentioned.
   Reference to fire illustrates the thoughtless and dangerous custom of whalers burning wrecked ships.
9. The suggestion seems to be that scurvy was beginning to affect the crew of the *Grenville Bay* as well. The next day's entry has the same flavour.
10. Cape Dyer on Baffin Island, not far north of the Arctic Circle.
11. Miss Peterkin's text says "fore and main sails." *The Orcadian* version seems much more likely.
12. 'Maintop' in *Orcadian* text.